Disclaimer

Book Title: Cell Phone Forensic Tools: An Overview and Analysis Update

Book Author: Richard P. Ayers; Wayne Jansen; Aurelien M. Delaitre; Ludovic Moenner

Book Abstract: Cell phones and other handheld devices incorporating cell phone capabilities (e.g., Personal Digital Assistant (PDA) phones) are ubiquitous. Rather than just placing calls, certain phones allow users to perform additional tasks such as SMS (Short Message Service) messaging, Multi-Media Messaging Service (MMS) messaging, IM (Instant Messaging), electronic mail, Web browsing, and basic PIM (Personal Information Management) applications (e.g., phone and date book). PDA phones, often referred to as smart phones, provide users with the combined capabilities of both a cell phone and a PDA. In addition to network services and basic PIM applications, one can manage more extensive appointment and contact information, review electronic documents, give a presentation, and perform other tasks. All but the most basic phones provide individuals with some ability to load additional applications, store and process personal and sensitive information independently of a desktop or notebook computer, and optionally synchronize the results at some later time. As digital technology evolves, the capabilities of these devices continue to improve rapidly. When cell phones or other cellular devices are involved in a crime or other incident, forensic examiners require tools that allow the proper retrieval and speedy examination of information present on the device. This report provides an overview on current tools (that have undergone significant updates or were not examined in NISTIR 7250: Cell Phone Forensic Tools: An Overview and Analysis) designed for acquisition, examination, and reporting of data discovered on cellular handheld devices, and an understanding of their capabilities and limitations.

Citation: NIST Interagency/Internal Report (NISTIR) - 7387

Keyword: cell phones; computer forensics; handheld devices; mobile devices

NIST

**National Institute of
Standards and Technology**

Technology Administration
U.S. Department of Commerce

NISTIR 7250

Cell Phone Forensic Tools:
An Overview and Analysis

Rick Ayers
Wayne Jansen
Nicolas Cilleros
Ronan Daniellou

NISTIR 7250

Cell Phone Forensic Tools:
An Overview and Analysis

Rick Ayers
Wayne Jansen
Nicolas Cilleros
Ronan Daniellou

C O M P U T E R   S E C U R I T Y

Computer Security Division
Information Technology Laboratory
National Institute of Standards and Technology
Gaithersburg, MD 20988-8930

October 2005

U.S. Department of Commerce
Carlos M. Gutierrez, Secretary

Technology Administration
Michelle O'Neill, Acting Under Secretary of
Commerce for Technology

National Institute of Standards and Technology
William A. Jeffrey, Director

Reports on Computer Systems Technology

The Information Technology Laboratory (ITL) at the National Institute of Standards and Technology (NIST) promotes the U.S. economy and public welfare by providing technical leadership for the Nation's measurement and standards infrastructure. ITL develops tests, test methods, reference data, proof of concept implementations, and technical analysis to advance the development and productive use of information technology. ITL's responsibilities include the development of technical, physical, administrative, and management standards and guidelines for the cost-effective security and privacy of sensitive unclassified information in Federal computer systems. This Interagency Report discusses ITL's research, guidance, and outreach efforts in computer security, and its collaborative activities with industry, government, and academic organizations.

National Institute of Standards and Technology Interagency Report
187 pages (2005)

Abstract

Cell phones and other handheld devices incorporating cell phone capabilities (e.g., Personal Digital Assistant (PDA) phones) are ubiquitous. Rather than just placing calls, certain phones allow users to perform additional tasks such as SMS (Short Message Service) messaging, Multi-Media Messaging Service (MMS) messaging, IM (Instant Messaging), electronic mail, Web browsing, and basic PIM (Personal Information Management) applications (e.g., phone and date book). PDA phones, often referred to as smart phones, provide users with the combined capabilities of both a cell phone and a PDA. In addition to network services and basic PIM applications, one can manage more extensive appointment and contact information, review electronic documents, give a presentation, and perform other tasks.

All but the most basic phones provide individuals with some ability to load additional applications, store and process personal and sensitive information independently of a desktop or notebook computer, and optionally synchronize the results at some later time. As digital technology evolves, the capabilities of these devices continue to improve rapidly. When cell phones or other cellular devices are involved in a crime or other incident, forensic examiners require tools that allow the proper retrieval and speedy examination of information present on the device. This report gives an overview of current forensic software, designed for acquisition, examination, and reporting of data discovered on cellular handheld devices, and an understanding of their capabilities and limitations.

Purpose and Scope

The purpose of this report is to inform law enforcement, incident response team members, and forensic examiners about the capabilities of present day forensic software tools that have the ability to acquire information from cell phones operating over CDMA (Code Division Multiple Access), TDMA (Time Division Multiple Access), GSM (Global System for Mobile communications) networks and running various operating systems, including Symbian, Research in Motion (RIM), Palm OS, Pocket PC, and Linux.

An overview of each tool describes the functional range and facilities for acquiring and analyzing evidence contained on cell phones and PDA phones. Generic scenarios were devised to mirror situations that arise during a forensic examination of these devices and their associated media. The scenarios are structured to reveal how selected tools react under various situations. Though generic scenarios were used in analyzing forensic tools, the procedures are not intended to serve as a formal product test or as a comprehensive evaluation. Additionally, no claims are made on the comparative benefits of one tool versus another. The report, instead, offers a broad and probing perspective on the state of the art of present-day forensic software tools for cell phones and PDA phones. Alternatives to using a forensic software tool for digital evidence recovery, such as desoldering and removing memory from a device to read out its contents or using a built-in hardware test interface to access memory, are outside the scope of this report.

It is important to distinguish this effort from the Computer Forensics Tool Testing (CFTT) project, whose objective is to provide measurable assurance to practitioners, researchers, and other users that the tools used in computer forensics investigations provide accurate results. Accomplishing this goal requires the development of rigorous specifications and test methods for computer forensics tools and the subsequent testing of specific tools against those specifications, which goes far beyond the analysis described in this document. The CFTT is the joint effort of the National Institute of Justice, the National Institute of Standards and Technology (NIST), the Office of Law Enforcement Standards (OLES), the U. S. Department of Defense, Federal Bureau of Investigation (FBI), U.S. Secret Service, the U.S. Immigration and Customs Enforcement (BICE), and other related agencies.[*]

The publication is not to be used as a step-by-step guide for executing a proper forensic investigation involving cell phones and PDA phones, or construed as legal advice. Its purpose is to inform readers of the various technologies available and areas for consideration when employing them. Before applying the material in this report, readers are advised to consult with management and legal officials for compliance with laws and regulations (i.e., local, state, federal, and international) that pertain to their situation.

---

[*] For more information on this effort see: **www.cftt.nist.gov**

Audience

The primary audience of the Cell Phone Forensic Tool document is law enforcement, incident response team members, and forensic examiners who are responsible for conducting forensic procedures related to cell phone devices and associated removable media.

Table of Contents

Acknowledgements

The authors, Rick Ayers, Wayne Jansen, Nicolas Cilleros, and Ronan Daniellou, wish to express their gratitude to colleagues who reviewed drafts of this document. In particular, their appreciation goes to Murugiah Souppaya, Karen Kent and Tim Grance from NIST, Karl Dunnagan from Mobile Forensics, Rick Mislan from Purdue University, and Eoghan Casey from Knowledge Solutions LLC and Brendan Farrah-Foley for their research, technical support, and written contributions to this document. The authors would also like to express thanks to all others who assisted with our internal review process, including Susan Ballou from NIST's Office of Law Enforcement Standards and Keith Thomas from Sytex, Inc.

This report was sponsored by Dr. Bert Coursey of the Department of Homeland Security (DHS). The Department's support and guidance in this effort are greatly appreciated.

Introduction

Computer forensics involves the identification, preservation, extraction, documentation, and analysis of computer data. Computer forensic examiners follow clear, well-defined methodologies and procedures that can be adapted for specific situations. Such methodologies consist of the following steps:

- Prepare a forensic copy (i.e., an identical bit-for-bit physical copy) of the acquired digital media, while preserving the acquired media's integrity.
- Examine the forensic copy to recover information.
- Analyze the recovered information and develop a report documenting any pertinent information uncovered.

Forensic toolkits are intended to facilitate the work of examiners, allowing them to perform the above steps in a timely and structured manner, and improve the quality of the results. This paper discusses available forensic software tools for handheld cellular devices, highlighting the facilities offered and associated capabilities.

Forensic software tools strive to address a wide range of applicable devices and handle the most common investigative situations with modest skill level requirements. These tools typically perform logical acquisitions using common protocols for synchronization, debugging, and communications. More complicated situations, such as the recovery of deleted data, often require highly specialized hardware-based tools and expertise, which is not within the scope of this report.

Handheld device forensics is a fairly new and emerging subject area within the computer forensics field, which traditionally emphasized individual workstations and network servers. Discrepancies between handheld device forensics and classical computer forensics exist due to several factors, including the following, which constrain the way in which the tools operate:

- The orientation toward mobility (e.g., compact size and battery powered, requiring specialized interfaces, media, and hardware)
- The filesystem residing in volatile memory versus non-volatile memory on certain systems
- Hibernation behavior, suspending processes when powered off or idle, but remaining active
- The diverse variety of embedded operating systems used
- Short product cycles for new handheld devices

Most cell phones offer a set of basic capabilities that are comparable. However, the various families of devices on the marketplace differ in such areas as the hardware technology, advanced feature set, and physical format. This paper looks at forensic software tools for a number of popular platforms, including Symbian, RIM (Research In Motion), Pocket PC, and Palm OS devices. Together these platforms comprise the majority of the so-called smart phone devices currently available and in use. More basic phones, produced by various manufacturers and operational on various types of cellular networks are also addressed in the paper.

The remaining sections provide an overview of cell phones, memory cards, and forensic toolkits; describe the scenarios used to analyze the various tools and toolkits; present the findings from applying the scenarios; and summarize the conclusions drawn. The reader is assumed to have some background in computer forensics and technology. The reader should also be apprised that the tools discussed in the report are rapidly evolving, with new versions and better capabilities available regularly. The tool manufacturer should always be contacted for up-to-date information.

Background

Cell phones are highly mobile communications devices that can do an array of functions ranging from that of a simple digital organizer to that of a low-end personal computer. Designed for mobility, they are compact in size, battery powered, and lightweight, often use proprietary interfaces or operating systems, and may have unique hardware characteristics for product differentiation. Overall, they can be classified as basic phones that are primarily simple voice and messaging communication devices; advanced phones that offer additional capabilities and services for multimedia; and smart phones or high-end phones that merge the capabilities of an advanced phone with those of a PDA (Personal Digital Assistant).

Figure 1 gives an overview of the hardware characteristics of basic, advanced, and high-end cell phones for display quality, processing and storage capacity, memory and I/O expansion, built-in communications, and video and image capture. The bottom of the diagram shows the range of cellular voice and data advances from kilobit analog networks, still in use today, to megabit 3rd generation digital networks in the planning and early deployment stages. The diagram attempts to illustrate that more capable phones can capture and retain not only more information, but also more varied information, through a wider variety of sources, including removable memory modules, other wireless interfaces, and built-in hardware. Note that hardware components can and do vary from those assignments made in the diagram and, over time, technology once considered high end or advanced eventually appears in what would then be considered a basic phone. Nevertheless, the principle remains true.

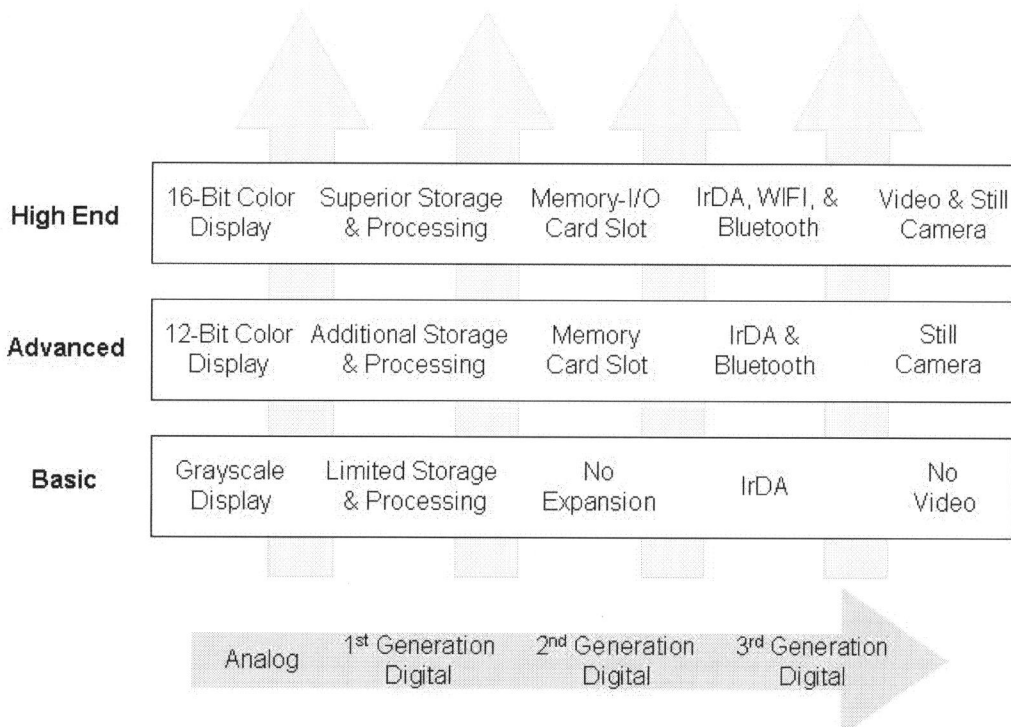

High End	16-Bit Color Display	Superior Storage & Processing	Memory-I/O Card Slot	IrDA, WIFI, & Bluetooth	Video & Still Camera
Advanced	12-Bit Color Display	Additional Storage & Processing	Memory Card Slot	IrDA & Bluetooth	Still Camera
Basic	Grayscale Display	Limited Storage & Processing	No Expansion	IrDA	No Video

Analog	1st Generation Digital	2nd Generation Digital	3rd Generation Digital

Figure 1: Phone Hardware Components

Just as with hardware components, software components involved in communications vary with the class of phone. Basic phones normally include text messaging using the Short Message Service (SMS). An advanced phone might add the ability to send simple picture messages or lengthy text messages using the Extended Message Service (EMS), while a high-end phone typically supports the Multimedia Message Service (MMS) to exchange sounds, color images, and text. Similarly, the ability to chat on-line directly with another user may be unsupported, supported through a dedicated SMS channel, or supported with a full Instant Messaging (IM) client. High-end phones typically support full function email and Web clients that respectively use POP (Post Office Protocol)/IMAP (Internet Message Access Protocol)/SMTP (Simple Mail Transfer Protocol) and HTTP, while advanced phones provide those services via WAP (Wireless Application Protocol), and basic phones do not include any support. Figure 2 gives an overview of the capabilities usually associated with each class of phone.

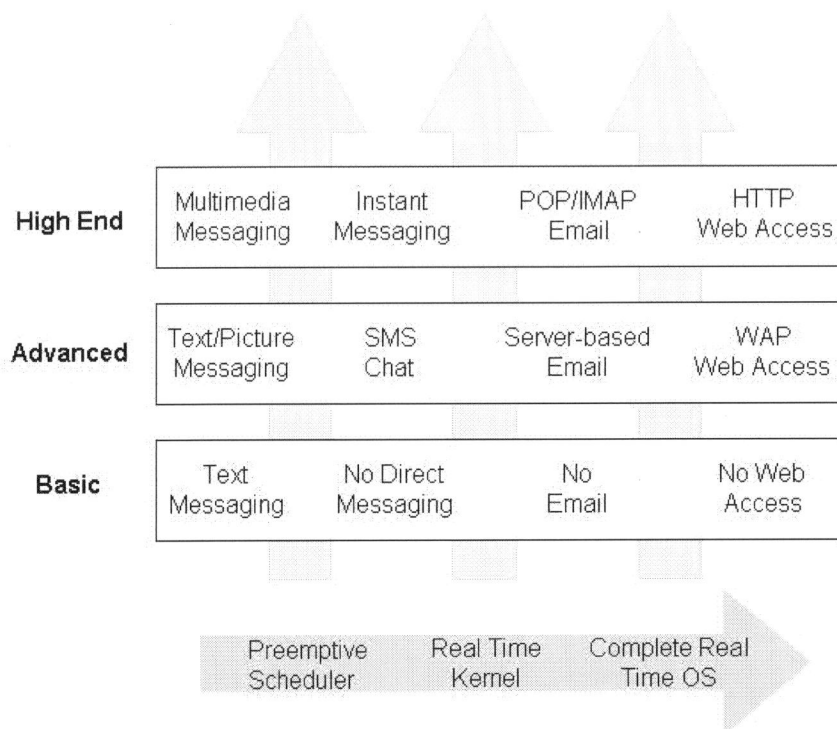

Figure 2: Phone Software Components

Most basic and many advanced phones rely on proprietary real-time operating systems developed by the manufacturer. Commercially embedded operating systems for cellular devices are also available that range from a basic preemptive scheduler with support for a few other key system calls to more sophisticated kernels with scheduling alternatives, memory management support, device drivers, and exception handling, to complete embedded real-time operating systems. The bottom of Figure 2 illustrates this range.

Many high-end smart phones have a PDA heritage, evolving from Palm OS and Pocket PC (also known as Windows mobile) handheld devices. As wireless telephony modules were incorporated into such devices, the operating system capabilities were enhanced to accommodate the functionality. Similarly, the Symbian OS found on many smart phones also stems from an electronic organizer heritage. RIM OS devices, which emphasize push technology for email messaging, are another device family that also falls into the smart phone category.

4

Subscriber Identity Module

Another useful way to classify cellular devices is by whether they involve a Subscriber Identity Module (SIM). A SIM is removable card designed for insertion into a device, such as a handset. SIMs originated with a set of specifications originally developed by the CEPT (Conference of European Posts and Telecommunications) and continued by ETSI (the European Telecommunications Standards Institute) for GSM networks. GSM standards mandate the use of a SIM for the operation of the phone. Without it, a GSM phone cannot operate. In contrast, present-day CDMA phones do not require a SIM. Instead, SIM functionality is incorporated directly within the device.

A SIM is an essential component of a GSM cell phone that contains information particular to the user. A SIM is a special type of smart card that typically contains between 16 to 64 KB of memory, a processor, and an operating system. A SIM uniquely identifies the subscriber, determines the phone's number, and contains the algorithms needed to authenticate a subscriber to a network. A user can remove the SIM from one phone, insert it into another compatible phone, and resume use without the need to involve the network operator. The hierarchically organized filesystem of a SIM is used to store names and phone numbers, received and sent text messages, and network configuration information. Depending on the phone, some of this information may also coexist in the memory of the phone or reside entirely in the memory of the phone instead of the SIM. While SIMs are most widely used in GSM systems, compatible modules are also used in IDEN phones and UMTS user equipment (i.e., a USIM). Because of the flexibility SIM offers GSM phone users to port their identity and information between devices, eventually all cellular phones are expected to include SIM capability.

Though two sizes of SIMs have been standardized, only the smaller size shown at left is broadly used in GSM phones today. The module has a width of 25 mm, a height of 15 mm, and a thickness of .76 mm, which is roughly the size of a postage stamp. Its 8-pin connectors are not aligned along a bottom edge as might be expected, but instead form a circular contact pad integral to the smart card chip, which is embedded in a plastic frame. Also, the slot for the SIM card is normally not accessible from the exterior of the phone as with a memory card. When a SIM is inserted into a phone and pin contact is made, a serial interface is used to communicate with the computing platform using a half-duplex protocol. SIMs can be removed from a phone and read using a specialized SIM card reader and software. A SIM can also be placed in a standard-size smart card adapter and read using a conventional smart card reader.

As with any smart card, its contents are protected and a PIN can be set to restrict access. Two PINs exist, sometimes called PIN1 and PIN2 or CHV1 and CHV2. These PINs can be modified or disabled by the user. The SIM allows only a preset number of attempts, usually three, to enter the correct PIN before further attempts are blocked. Entering the correct PUK (PIN Unblocking Key) resets the PIN number and the attempt counter. The PUK can be obtained from the service provider or the network operator based on the SIM's identity (i.e., its ICCID). If the number of attempts to enter the PUK correctly exceeds a set limit, normally ten attempts, the card becomes blocked permanently.

Removable media extends the storage capacity of a cell phone, allowing individuals to store additional information beyond the device's built-in capacity. They also provide another avenue for sharing information between users that have compatible hardware. Removable media is non-volatile storage, able to retain recorded data when removed from a device. The main type of removable media for cell phones is a memory card. Though similar to a SIM in size, they follow a different set of specifications and have vastly different characteristics. Some card specifications also allow for I/O capabilities to support wireless communications (e.g., Bluetooth or WiFi) or other hardware (e.g., a camera) to be packaged in the same format.

A wide array of memory cards exists on the market today for cell phones and other mobile devices. The storage capacities of memory cards range from megabytes (MB) to gigabytes (GB) and come in sizes literally as small as a thumbnail. As technological advances continue, such media is expected to become smaller and offer greater storage densities. Fortunately, such media is normally formatted with a conventional filesystem (e.g., FAT) and can be treated similarly to a disk drive, imaged and analyzed using a conventional forensic tool with a compatible media adapter that supports an Integrated Development Environment (IDE) interface. Such adapters can be used with a write blocker to ensure that the contents remain unaltered. Below is a brief overview of several commonly available types of memory cards used with cell phones.

Multi-Media Cards (MMC):[1]

A Multi-Media Card (MMC) is a solid-state disk card with a 7-pin connector. MMC cards have a 1-bit data bus. They are designed with flash technology, a non-volatile storage technology that retains information once power is removed from the card. Multi-Media Cards are about the size of a postage stamp (length-32 mm, width-24 mm, and thickness-1.4 mm). Reduced Size Multi-Media cards (RS-MMC) also exist. They are approximately one-half the size of the standard MMC card (length-18mm, width-24mm, and thickness-1.4mm). An RS-MMC can be used in a full-size MMC slot with a mechanical adapter. A regular MMC card can be also used in an RS-MMC card slot, though part of it will stick out from the slot. MMCplus and MMCmobile are higher performance variants of MMC and RS-MMC cards respectively that have a 13-pin connector and an 8-bit data bus.

Secure Digital (SD) Cards:[2]

Secure Digital (SD) memory cards (length-32 mm, width-24 mm, and thickness-2.1mm) are comparable to the size and solid-state design of MMC cards. In fact, SD card slots can often accommodate MMC cards as well. However, SD cards have a 9-pin connector and a 4-bit data bus, which afford a higher transfer rate. SD memory cards feature an erasure-prevention switch. Keeping the switch in the locked position protects data from accidental deletion. They also offer security controls for content protection (i.e., Content Protection Rights Management). MiniSD cards are an electrically compatible extension of the existing SD card standard in a more compact format (length-21.5 mm, width-20 mm, and thickness-1.4 mm). They run on the same hardware bus and use the same interface as an SD card, and also include content protection security features, but have a smaller maximum capacity

---

[1] Image courtesy of Lexar Media. Used by permission.
[2] Image courtesy of Lexar Media. Used by permission.

potential due to size limitations. For backward compatibility, an adapter allows a MiniSD Card to work with existing SD card slots.

Memory Sticks:[3]

Memory sticks provide solid-state memory in a size similar to, but smaller than, a stick of gum (length-50mm, width-21.45mm, thickness-2.8mm). They have a 10-pin connector and a 1-bit data bus. As with SD cards, memory sticks also have a built-in erasure-prevention switch to protect the contents of the card. Memory Stick PRO cards offer higher capacity and transfer rates than standard Memory Sticks, using a 10-pin connector, but with a 4-bit data bus. Memory Stick Duo and Memory Stick PRO Duo, smaller versions of the Memory Stick and Memory Stick PRO, are about two-thirds the size of the standard memory stick (length-31mm, width-20mm, thickness-1.6mm). An adapter is required for a Memory Stick Duo or a Memory Stick PRO Duo to work with standard Memory Stick slots.

TransFlash:[4]

TransFlash is a tiny memory card based on the MiniSD card. Because of their extremely small size (length-15 mm, width-11 mm, and thickness-1 mm), frequent removal and handling is discouraged, making them more of a semi-removable memory module. TransFlash cards have an 8-pin connector and a 4-bit data bus. An adapter allows a TransFlash card to be used in SD-enabled devices. Similarly, the newly announced MMCmicro device is another ultra small card (length-14 mm, width-12 mm, and thickness-1.1 mm), compatible with MMC-enabled devices via an adapter. MMCmicro cards have a 10-pin connector and a 1 or 4-bit data bus. TransFlash has recently been renamed MicroSD.

---

[3] Image courtesy of Lexar Media. Used by permission.
[4] Image courtesy of SanDisk. Used by permission.

Forensic Toolkits

The variety of forensic toolkits for cell phones and other handheld devices is diverse. A considerable number of software tools and toolkits exist, but the range of devices over which they operate is typically narrowed to distinct platforms for a manufacturer's product line, a family of operating systems, or a type of hardware architecture. Moreover, the tools require that the examiner have full access to the device (i.e., the device is not protected by some authentication mechanism or the examiner can satisfy any authentication mechanism encountered).

While most toolkits support a full range of acquisition, examination, and reporting functions, some tools focus on a subset. Similarly, different tools may be capable of using different interfaces (e.g., IrDA, Bluetooth, or serial cable) to acquire device contents. The types of information that tool can acquire can range widely and include PIM (Personal Information Management) data (e.g., phone book); logs of phone calls; SMS/EMS/MMS messages, email, and IM content; URLs and content of visited Web sites; audio, video, and image content; SIM content; and uninterrupted image data. Information present on a cell phone can vary depending on several factors, including the following:
- The inherent capabilities of the phone implemented by the manufacturer
- The modifications made to the phone by the service provider or network operator
- The network services subscribed to and used by the user
- The modifications made to the phone by the user

Acquisition through a cable interface generally yields superior acquisition results than other device interfaces. However, a wireless interface such as infrared or Bluetooth can serve as a reasonable alternative when the correct cable is not readily available. Regardless of the interface used, one must be vigilant about any forensic issues associated. Note too that the ability to acquire the contents of a resident SIM may not be supported by some tools, particularly those strongly oriented toward PDAs. Table 1 lists open-source and commercially available tools and the facilities they provide for certain types of cell phones.

Table 1: Cell Phone Tools

	Function	Features
PDA Seizure	Acquisition, Examination, Reporting	• Targets Palm OS, Pocket PC, and RIM OS phones • No support for recovering SIM information • Supports only cable interface
pilot-link	Acquisition	• Targets Palm OS phones • Open source non-forensic software • No support for recovering SIM information • Supports only cable interface
Cell Seizure	Acquisition, Examination, Reporting	• Targets certain models of GSM, TDMA, and CDMA phones • Supports recovery of internal and external SIM • Supports only cable interface

8

	Function	Features
GSM .XRY	Acquisition, Examination, Reporting	• Targets certain models of GSM phones • Supports recovery of internal and external SIM • Supports cable, Bluetooth, and IR interfaces
Oxygen PM (forensic version)	Acquisition, Examination, Reporting	• Targets certain models of GSM phones • Supports only internal SIM acquisition
MOBILedit! Forensic	Acquisition, Examination, Reporting	• Targets certain models of GSM phones • Internal and external SIM support • Supports cable and IR interfaces
BitPIM	Acquisition, Examination	• Targets certain models of CDMA phones • Open source software with write-blocking capabilities • No support for recovering SIM information
TULP 2G	Acquisition, Reporting	• Targets GSM and CDMA phones that use the supported protocols to establish connectivity • Internal and external SIM support • Requires PC/SC-compatible smart card reader for external SIM cards • Cable, Bluetooth, and IR interfaces supported

Because of the way GSM phones are logically and physically partitioned into a handset and SIM, a number of forensic software tools have emerged that deal exclusively with SIMs independently of their handsets. The SIM must be removed from the phone and inserted into an appropriate reader for acquisition. SIM forensic tools require either a specialized reader that accepts a SIM directly or a general-purpose reader for a full-size smart card. For the latter, a standard-size smart card adapter is needed to house the SIM for use with the reader. Table 2 lists several SIM forensic tools. The first four listed, Cell Seizure, TULP2G, GSM .XRY, and Mobiledit!, also handle phone memory acquisition, as noted above.

Table 2: SIM Tools

	Function	Features
Cell Seizure	Acquisition, Examination, Reporting	• Also recover information from SIM card, when inserted in handset
TULP 2G	Acquisition, Reporting	• Also recover information from SIM card, when inserted in handset
GSM .XRY	Acquisition, Examination, Reporting	• Also recover information from SIM card, when inserted in handset
Mobiledit! Forensic	Acquisition, Examination, Reporting	• Also recover information from SIM card, when inserted in handset
SIMIS	Acquisition, Examination, Reporting	• External SIM cards only

	Function	Features
ForensicSIM	Acquisition, Examination, Reporting	• External SIM cards only • Produces physical facsimiles of SIM for prosecutor and defense, and as a storage record
Forensic Card Reader	Acquisition, Reporting	• External SIM cards only
SIMCon	Acquisition, Examination, Reporting	• External SIM cards only

Forensic software tools acquire data from a device in one of two ways: physical acquisition or logical acquisition. Physical acquisition implies a bit-by-bit copy of an entire physical store (e.g., a disk drive or RAM chip), while logical acquisition implies a bit-by-bit copy of logical storage objects (e.g., directories and files) that reside on a logical store. The difference lies in the distinction between memory as seen by a process through the operating system facilities (i.e., a logical view), versus memory as seen by the processor and other hardware components (i.e., a physical view). In general, physical acquisition is preferable, since it allows any data remnants present (e.g., unallocated RAM or unused filesystem space) to be examined, which otherwise would go unaccounted in a logical acquisition. Physical device images are generally more easily imported into another tool for examination and reporting. However, a logical acquisition provides a more natural and understandable organization of the information acquired. Thus, if possible, doing both types of acquisition is preferable.

Tools not designed specifically for forensic purposes are questionable and should be thoroughly evaluated before use. Though both forensic and non-forensic software tools generally use the same protocols to communicate with the device, non-forensic tools allow a two-way flow of information in order to populate and manage the device, and avoid taking hashes of acquired content for integrity purposes. Documentation also may be limited and source code unavailable for examination, respectively increasing the likelihood of error and decreasing confidence in the results. On the one hand, non-forensic tools might be the only means to retrieve information that could be relevant as evidence. On the other, they might overwrite, append, or otherwise cause information to be lost, if not used carefully.

The remainder of this chapter provides a brief introduction to each tool used for this report.

PDA Seizure

Paraben's PDA Seizure version 3.0.3.89[5] is a forensic software toolkit that allows forensic examiners to acquire, search, examine, and report data associated with PDAs running Palm OS, Windows CE, and RIM OS. Though able to be used with smart phones running these operating systems, the toolkit is oriented toward non-cellular devices and omits cell phone-oriented features, such as SIM acquisition for GSM phones. PDA Seizure's features include the ability to perform a logical and physical acquisition, providing a view of internal memory and relevant information concerning individual files and databases. PDA Seizure uses the MD5 hash function

---

[5] Additional information on Paraben products can be found at: **http://www.paraben-forensics.com**

to protect the integrity of acquired files. Additional features include bookmarking of information to be filtered and organized in a report format, searching for text strings within the acquired data, and automatic assembly found images under a single facility.

Pilot-Link

pilot-link[6] is an open source software suite originally developed for the Linux community to allow information to be transferred between Linux hosts and Palm OS devices. It runs on several other desktop operating systems besides Linux, including Windows and Mac OS. About thirty command line programs comprise the software suite. To perform a physical and logical dump, pilot-link establishes a connection to the device with the aid of the Hotsync protocol. The two programs of interest to forensic examiners are pi-getram and pi-getrom, which respectively retrieve the physical contents of RAM and ROM from a device. Another useful program is pilot-xfer, which allows the installation of programs and the backup and restoration of databases. pilot-xfer provides a means to acquire the contents of a device logically. The contents retrieved with these utilities can be manually examined with either the Palm OS Emulator (POSE), a compatible forensics tool such as EnCase, or a hex editor. pilot-link does not provide hash values of the information acquired. A separate step must be carried out to obtain needed hash values.

Cell Seizure

Paraben's Cell Seizure version 2.0.0.33660[7] is a forensic software toolkit that allows forensic examiners to acquire, search, examine, and report data associated with cell phones operating over CDMA, TDMA, and GSM networks. To acquire data from cell phones using Paraben's Cell Seizure software, the proper cable must be selected from either the Cell Seizure Toolbox or a compatible cable (e.g., datapilot) to establish a data-link between the phone and the forensic workstation. The type of phone being acquired determines the cable interface. Serial RS-232 and USB data-link connections are established via the phone data port or the under-battery interface connection. Additional features include bookmarking of information to be filtered and organized in a format report, searching for case-sensitive whole word text and hexadecimal values, and automatic assembly of found images under a single facility. The following data can usually be found on most cell phones with the tool:

- SMS History: Inbox/Outbox
- Phonebook: SIM-Card, Own Numbers, Speed Dialing, Fixed Dialing
- Call Logs: Dialed Numbers, Received Calls, Missed Calls
- Calendar: Reminder, Meeting, Memo
- Logos: Caller Logos, Startup Logos, Welcome Notes
- Graphics: Wallpaper, Picture Camera Images, EMS Template Images
- WAP: WAP Settings, WAP Bookmarks
- SIM: GSM Specific data

---

[6] Additional information on pilot-link can be found at: **http://www.pilot-link.org**
[7] Additional information on Paraben products can be found at: **http://www.paraben-forensics.com**

GSM .XRY

Micro Systemation's SoftGSM .XRY is a forensic software toolkit for acquiring data from GSM, CDMA, 3G phones and SIM/USIM cards. The .XRY unit is able to connect to cell phone devices via Infrared (IR) port, Bluetooth or a cable interface. After establishing connectivity, the phone model is identified with a corresponding picture of the phone, the device name, manufacturer, model, serial number (IMEI), Subscriber ID (IMSI), manufacturer code, device clock, and the PC clock. Data acquired from cell phone devices are stored in the .XRY format and cannot be altered, but can be exported into external formats and viewed with third-party applications. After a successful acquisition, the following fields may be populated with data, depending on the phone's functionality: Summary screen, Case data, General Information, Contacts, Calls, Calendar, SMS, Pictures, Audio, Files, Notes, Tasks, MMS, Network Information, Video, etc. Additionally, graphic files, audio files, and internal files present on the phone can be viewed internally or exported to the forensic workstation for safekeeping or further investigation.

Oxygen Phone Manager

The forensic version of Oxygen Phone Manager is available for Police Departments, Law Enforcement units, and all government services that wish to use the software for investigation purposes. The forensic version differs from the non-forensic version of Oxygen Phone Manager by prohibiting any changes in data during acquisition. Oxygen Phone Manager (OPM) allows examiners to acquire data from the device and export the acquired data into multiple supported formats. The Oxygen software is tailored toward mobile phones and smart phones manufactured by: Nokia, Sony Ericsson, Siemens, Panasonic, Sendo, BenQ and some Samsung models. Oxygen software provides software libraries, ActiveX libraries and components for Borland Delphi to software developers.

MOBILedit!

MOBILedit! Forensic is an application giving examiners the ability to acquire logically, search, examine and report data from GSM/CDMA/PCS cell phone devices. MOBILedit! is able to connect to cell phone devices via an Infrared (IR) port, a Bluetooth link, or a cable interface. After connectivity has been established, the phone model is identified by its manufacturer, model number, and serial number (IMEI) and with a corresponding picture of the phone. Data acquired from cell phone devices are stored in the .med file format. After a successful acquisition, the following fields are populated with data: subscriber information, device specifics, Phonebook, SIM Phonebook, Missed Calls, Last Numbers Dialed, Received Calls, Inbox, Sent Items, Drafts, Files folder. Items present in the Files folder, ranging from Graphics files to Camera Photos and Tones, depend on the phone's capabilities. Additional features include the myPhoneSafe.com service, which provides access to the IMEI database to register and check for stolen phones.

BitPIM

BitPIM is a phone management program that runs on Windows, Linux and Mac OS and allows the viewing and manipulation of data on cell phones. This data includes the phone book, calendar, wallpapers, ring tones and the embedded filesystem. To acquire data successfully using BitPIM, examiners must have the proper driver and cable to form a connection between the phone and the forensic workstation. BitPIM provides detailed information contained in the help file, outlining supported phones, suggested cables to use with specific phone models, and notes and How-Tos about specific situations. BitPIM is distributed as open source software under the GNU General Public License.

TULP 2G

TULP2G (2nd generation) is an open source, forensic software tool originated by the Netherlands Forensic Institute that allows examiners to extract and read data from mobile cell phones and SIMs. TULP2G requires a forensic workstation running either Windows 2000 or XP, preferably with the latest patches and service pack installed, along with .NET 1.1 SP1. In order to take advantage of newly released 1.1 plug-ins, Windows XP SP2 is required. TULP2G acquires data from mobile phones using a proper data cable, Bluetooth or IrDA connection and a compatible protocol plug-in. Reading SIMs requires a PC/SC-compatible smart card reader and possibly an adapter to convert a small-sized SIM to the standard-size smart card format.

SIMIS

SIMIS is a forensic tool from Crownhill USA[8] that allows examiners the ability to extract data from a SIM securely and protect the integrity with cryptographic hashes. A USB dongle is needed to operate the software on a desktop computer. The SIMIS desktop is capable of decoding unicode data found on the SIM Card, including active and deleted text messages and phone book information. The company also offers the SIMIS Mobile Handheld Reader, which is a portable stand-alone SIM reader that can capture SIM data for transfer to the SIMIS desktop.

ForensicSIM

Radio Tactic's ForensicSIM Toolkit consists of the following components: acquisition terminal, control card, data storage cards, analysis application, and the card reader. The acquisition terminal is a stand-alone unit that guides the examiner through each step of the acquisition process. The ForensicSIM toolkit deals with two processes: acquisition of data and analysis of data. Data acquisition is carried out using the acquisition terminal. Data analysis is carried out using the ForensicSIM card reader, attached to a PC running the ForensicSIM analysis application. The terminal's primary function is to capture copies of the data from the target SIM to a set of data storage cards. A control card is used to provide the examiner access to the acquisition terminal, thwarting unauthorized use. The data storage cards consist of a master data storage card, a prosecution data storage card, a defense data storage card, and a handset access card. The toolkit allows examiners read-only access to SIMs and generates textual reports based on the contents acquired. Reports can be viewed internally, saved to disk, or printed for presentation purposes.

Forensic Card Reader

The Forensic Card Reader (FCR) consists of a smart card reader with USB connection and the FCR software that gives examiners the ability to acquire data from SIM cards without modification. The examiner has the ability to select specific data elements that can be later stored and displayed in a finalized report. Operations details like case number, evidence number, and examiner can be automatically merged into the report and its file name. All usual data elements are acquired (e.g., phone directory, abbreviated dialing numbers, fixed dialing numbers

---

[8] For additional information on Crownhill USA products see: **www.crownhillmobile.com**

13

and SMS messages), as well as the identifiers of the SIM and the subscriber. Special elements such as deleted SMS messages can also be acquired. The FCR stores a complete report in an XML format. SIM cards for GSM mobiles and also SIM cards for 3G mobiles can be used with the FCR. Extended phone book entries can be acquired, including additional numbers and email addresses. The supplied FCR reader allows examiners to use either small or large SIM cards without the need for an adapter.

SIMCon

SIMCon works with any standard smart card reader compliant with the PC/SC standard. Upon completing the acquisition of the SIM card data, SIMCon card content is stored in unique files identified by a two-byte File ID code. Individual files may contain many informative elements called "items" and are displayed in tabular form. Each item, when selected, can be shown in hexadecimal or a textual interpretation. Besides standard SIM file content, SIMCon also has an option to do a comprehensive scan of all directories and files that may be present on the SIM, to acquire non-standardized directories and files. Examiners can create customized reports by selecting file information that pertains to the investigation.

Analysis Overview

A simple methodology was followed to understand and gauge the capabilities of the forensic tools described in the previous section. The main steps are illustrated in Figure 3. First, a set of target devices ranging from simple to smart phones was assembled. Then, a set of prescribed activities, such as placing and receiving calls, was performed for each phone. After each such scenario, the contents of the phone and/or associated SIM were acquired using an available tool and examined to see if the results of an activity could be recovered as expected. Finally, an assignment was made about how well the tool met predefined expectations. The process was repeated for each scenario defined. At least two different individuals performed each scenario and assigned a rating separately; any noted inconsistencies were resolved.

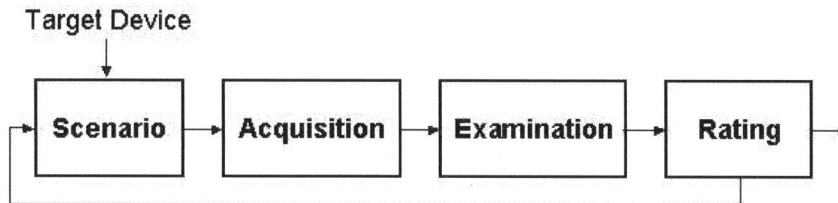

Figure 3: Tool Assessment

For GSM phones, two sets of scenarios were applied: one for handsets containing an associated SIM, and the other for SIMs removed from their handsets and examined independently. For CDMA and other types of phones that do not depend on a SIM, only the former set was used.

Target Devices

A suitable but limited number of target devices were needed on which to conduct the scenarios. The target devices selected, while not extensive, cover a range of operating systems, processor types, and hardware components. These variations were intended to uncover subtle differences in the behavior of the forensic tools in acquisition and examination. Table 3 highlights the key characteristics of each target device, listed roughly from devices with more capabilities to less-capable devices, rather than alphabetically. Note that the more capable devices listed have a PDA heritage, insofar as they use Windows Mobile, Palm OS, RIM OS, and Symbian operating systems.

Table 3: Target Device Characteristics

	Software	Hardware	Wireless
Motorola MPX220	Windows Mobile for Smart phones 2003 SMS, EMS, MMS SMS Chat Email (IMAP4, POP3) Web (HTML, WAP 2.0)	200 MHz OMAP 1611 processor 64 MB ROM 32 MB RAM Color display 2nd monochrome display Camera MiniSD slot	GSM 850/900/ 1800/1900 GPRS Bluetooth IrDA

15

	Software	Hardware	Wireless
Treo 600	Palm OS 5.2 SMS, EMS, MMS SMS Chat Email (POP3, SMTP) Web (HTML 4.0, XHTML, WML 1.3)	144 MHz OMAP 1510 ARM-based processor 32 megs of RAM (24 MB available) Color display QUERTY keypad SD/MMC slot (with SDIO)	GSM 850/900/ 1800/1900 GPRS IrDA
Sony Ericsson P910a	Symbian 7.0, UIQ 2.1 SMS, EMS, MMS Email (POP3, IMAP4) Web (WAP)	ARM 9 processor 64MB ROM 32MB RAM Color display Camera Memory Stick duo pro slot	GSM 850/1800/1900 HSCSD, GPRS Bluetooth IrDA
Samsung i700	Pocket PC 2002 Phone Edition SMS (no EMS/MMS) Email Web Instant Messaging	300 MHz StrongArm PXA250 processor 32MB flash memory 64MB SDRAM Color display Swivel camera SD/MMC slot (with SDIO)	AMPS 800 CDMA 800/1900 1xRTT IrDA
Nokia 7610	Symbian 7.0, Series 60 2.0 SMS, MMSConcatenated SMS Email (SMTP, POP3, IMAP4) Instant Messaging Web (WAP 2.0, HTML, XHTML and WML)	123 MHz processor 8 MB internal dynamic memory Color display Camera Reduced Size MMC slot	GSM 850/1800/1900 HSCSD, GPRS Bluetooth
Kyocera 7135	Palm OS 4.1 SMS, EMS (no MMS) Email (POP, IMAP, SMTP) Web (HTML 3.2)	33 MHz Dragonball VZ processor 16 MB volatile Color Display SD/MMC slot (with SDIO)	AMPS 800 CDMA 800/1900 1xRTT IrDA
BlackBerry 7780	RIM OS SMS Email (POP3) Web (WAP)	16 MB flash memory plus 2 MB SRAM Color display QWERTY keypad	GSM 850/1800/1900 GPRS
BlackBerry 7750	RIM OS SMS (no EMS/SMS) Email (POP3, IMAP4) Web (WAP 2.0, WML/HTML)	ARM7TDMI (Qualcomm 5100 Chipset) 14 MB flash memory 2 MB SRAM Color display QWERTY keypad	CDMA 800/1900 1xRTT

	Software	Hardware	Wireless
Motorola V300	SMS, EMS, MMS SMS Chat Nokia Smart Message Instant Messaging Email (SMTP, POP3, IMAP4) Web (WAP 2.0)	Internal Memory 5MB Color display Camera	GSM 900/1800/1900 GPRS
Nokia 6610i	Series 40 SMS, MMS Concatenated SMS SMS Chat No Email Web (WAP 1.2.1 XHTML)	4 MB user memory 8-line color display Camera FM radio	GSM 900/1800/1900 HSCSD, GPRS IrDA
Ericsson T68i	SMS/EMS messaging MMS messaging Email (POP3,SMTP) SMS Chat Web (WAP 1.2.1/2.0, WLTS)	Color display Optional camera attachment	GSM 900/1800/1900 HSCSD, GPRS Bluetooth IrDA
Sanyo 8200	SMS, EMS Picture Mail Email Web WAP 2.0 Mobile-to-mobile (walkie talkie)	Color display 2nd color display Camera	AMPS 850 CDMA 850/1900
Nokia 6200	SMS, EMS, MMS Email over SMS SMS Chat Web (WAP 1.2.1, XHTML)	Color display FM radio	GSM 850/1800/1900 GPRS, EDGE IrDA
Audiovox 8910	EMS, MMS SMS Chat No email Web (WAP 2.0)	Color display 2nd monochrome display Camera	AMPS 850 CDMA800/1900 1xRTT
Motorola C333	SMS, EMS SMS chat WAP 1.2.1	Monochrome graphic display	GSM 850/1900 GPRS
Motorola V66	SMS (no EMS) AOL Instant Messenger Web (WAP 1.1)	Monochrome graphic display	GSM 900/1800/1900 GPRS
Nokia 3390	SMS Picture messaging Email over SMS AOL Instant Messaging	Monochrome graphic display	GSM 1900

Every tool does not support every target device. In fact, the opposite is true – a specific tool typically supports only a limited number of devices. The determination of which tool to use for which device was based primarily on the tool's documented list of supported phones. Whenever ambiguity existed, an acquisition attempt was conducted to make a determination. Table 4 summarizes the various target devices used with each tool. The order of the devices bears no relevance on capabilities they are alphabetized for consistency throughout the remaining portion of the document. The table excludes forensic SIM tools, which support most SIMs found in GSM devices.

Table 4: Target Devices Supported by Each Tool

	PDA Seizure	Pilot-link	Cell Seizure	GSM .XRY	OPM	MOBILedit!	TULP 2G	BITpim
Audiovox 8910								X
BlackBerry 7750	X							
BlackBerry 7780	X							
Ericsson T68i			X	X		X	X	
Kyocera 7135	X	X						
Motorola C333			X			X	X	
Motorola MPX220	X							
Motorola V66			X	X		X	X	
Motorola V300				X		X	X	
Nokia 3390			X		X			
Nokia 6610i			X	X	X	X	X	
Nokia 6200				X	X		X	
Nokia 7610				X	X			
Samsung i700	X							
Sanyo 8200								X

18

	PDA Seizure	Pilot-link	Cell Seizure	GSM .XRY	OPM	MOBILedit!	TULP 2G	BITpim
Sony Ericsson P910a							X	
Treo 600	X	X						

Though SIMs are highly standardized, their content can vary among network operators and service providers. For example, a network operator might create an additional file on the SIM for use in its operations or might install an application to provide a unique service. SIMs may also be classified according to the "phase" of the GSM standards that they support. The three phases defined are phase 1, phase 2, and phase 2+, which correspond roughly to first, second, and 2.5 generation network facilities. Another class of SIMs in early deployment is Universal SIMs (USIMS) used in third generation (3G) networks. Table 5 lists the identifier and phase of the SIMs used in the analysis, the associated network operator, and some of the associated network services activated on the SIM. Except for pay-as-you-go phones, each GSM phone was matched with a SIM that offered services compatible with the phone's capabilities.

Table 5: SIMs

SIM	Phase	Network	Services
1604	2 - profile download required	AT&T	Abbreviated Dialing Numbers (ADN) Fixed Dialing Numbers (FDN) Short Message Storage (SMS) Last Numbers Dialed (LND) Group Identifier Level 1 (GID1) Group Identifier Level 2 (GID2) Service Dialing Numbers (SDN) General Packet Radio Service (GPRS)
1144	2 - profile download required	AT&T	Abbreviated Dialing Numbers (ADN) Fixed Dialing Numbers (FDN) Short Message Storage (SMS) Last Numbers Dialed (LND) General Packet Radio Service (GPRS)
8778	2- profile download required	Cingular	Abbreviated Dialing Numbers (ADN) Fixed Dialing Numbers (FDN) Short Message Storage (SMS) Last Numbers Dialed (LND) Group Identifier Level 1 (GID1) Group Identifier Level 2 (GID2) Service Dialing Numbers (SDN) General Packet Radio Service (GPRS)

SIM	Phase	Network	Services
7202	2 - profile download required	T-Mobile	Abbreviated Dialing Numbers (ADN) Fixed Dialing Numbers (FDN) Short Message Storage (SMS) Last Numbers Dialed (LND) Group Identifier Level 1 (GID1) General Packet Radio Service (GPRS)
5343	2 - profile download required	T-Mobile	Abbreviated Dialing Numbers (ADN) Fixed Dialing Numbers (FDN) Short Message Storage (SMS) Last Numbers Dialed (LND) General Packet Radio Service (GPRS)

Overall, SIM forensic tools do not recover every possible item on a SIM. The breadth of coverage also varies considerably among tools. Table 6 entries give an overview of those items recovered, listed at the left, by the various SIM forensic tools, listed across the top.

Table 6: Content Recovery Coverage

	Cell Seizure	GSM .XRY	Mobiledit!	TULP 2G	FCR	Forensic SIM	SIMCon	SIMIS
International Mobile Subscriber Identity – IMSI	X	X	X	X	X	X	X	X
Integrated Circuit Card Identifier – ICCID	X	X	X	X	X	X	X	X
Mobile Subscriber ISDN -- MSISDN	X	X		X	X	X	X	X
Service Provider Name – SPN	X			X		X	X	X
Phase identification – Phase	X	X	X			X	X	X
SIM Service Table – SST				X		X	X	X
Language Preference – LP	X			X		X	X	X

	Cell Seizure	GSM .XRY	Mobiledit!	TULP 2G	FCR	Forensic SIM	SIMCon	SIMIS
Abbreviated Dialing Numbers – ADN	X	X	X	X	X	X	X	X
Last Numbers Dialed – LND	X	X	X	X	X	X	X	X
Short Message Service – SMS • *Read/Unread* • *Deleted*	X X	X X	X 	X X	X 	X X	X X	X X
PLMN selector – PLMNsel	X			X		X	X	X
Forbidden PLMNs – FPLMNs	X			X		X	X	X
Location Information – LOCI	X	X		X	X	X	X	X
GPRS Location Information – GPRSLOCI	X					X	X	X

A subset of the SIMs used for the phone scenarios were used in the SIM scenarios. Appendix Q: SIM Services provides a full list of the SIM services associated with each of these SIMs.

Scenarios

The scenarios define a set of prescribed activities used to gauge the capabilities of the forensic tool to recover information from a phone, beginning with connectivity and acquisition and moving progressively toward more interesting situations involving common applications, file formats, and device settings. The scenarios are not intended to be exhaustive or to serve as a formal product evaluation. However, they attempt to cover a range of situations commonly encountered when examining a device (e.g., data obfuscation, data hiding, data purging) and are useful in determining the features and functionality afforded an examiner.

Table 7 gives an overview of these scenarios, which are generic to all devices that have cellular phone capabilities. For each scenario listed, a description of its purpose, method of execution, and expected results are summarized. Note that the expectations are comparable to those an examiner would have when dealing with the contents of a hard disk drive as opposed to a PDA/cell phone. Though the characteristics of the two are quite different, the recovery and analysis of information from a hard drive is a well-understood baseline for comparison and pedagogical purposes. Moreover, comparable means of digital evidence recovery from most phones exists, such as desoldering and removing non-volatile memory and reading out the contents with a suitable device programmer. Also note that none of the scenarios attempt to confirm whether the integrity of the data on a device is preserved when applying a tool – that topic is outside the scope of this document.

Table 7: Phone Scenarios

Scenario	Description
Connectivity and Retrieval	Determine if the tool can successfully connect to the device and retrieve content from it. • Enable user authentication on the device before acquisition, requiring a PIN, password, or other known authentication information to be supplied for access. • Initiate the tool on a forensic workstation, attempt to connect with the device and acquire its contents, verify that the results are consistent with the known characteristics of the device. • Expect that the authentication mechanism(s) can be satisfied without affecting the tool, and information residing on the device can be retrieved.
PIM Applications	Determine whether the tool can find information, including deleted information, associated with Personal Information Management (PIM) applications such as phone book and date book. • Create various types of PIM files on the device, selectively delete some entries, acquire the contents of the device, locate and display the information. • Expect that all PIM-related information on the device can be found and reported, if not previously deleted. Expect that remnants of deleted information can be recovered and reported.
Dialed/Received Phone Calls	Determine whether the tool can find dialed and received phone calls, including unanswered and deleted calls. • Place and receive various calls to and from different numbers, selectively delete some entries, acquire the contents of the device, locate and display dialed and received calls. • Expect that all dialed and received phone calls on the device can be recognized and reported, if not previously deleted. Expect that remnants of deleted information can be recovered and reported.
SMS/MMS Messaging	Determine whether the tool can find placed and received SMS/MMS messages, including deleted messages. • Place and receive both SMS and MMS messages, selectively delete some messages, acquire the contents of the device, locate and display all messages. • Expect that all sent and received SMS/MMS messages on the device can be recognized and reported, if not previously deleted. Expect that remnants of deleted information can be recovered and reported.

Scenario	Description
Internet Messaging	Determine whether the tool can find sent and received email and Instant Message (IM) messages, including deleted messages. • Send and receive both IM and email messages, selectively delete some messages, acquire the contents of the device, locate and display all messages. • Expect that all sent and received IM and messages on the device can be recognized and reported, if not previously deleted. Expect that remnants of deleted information can be recovered and reported.
Web Applications	Determine whether the tool can find a visited Web site and information exchanged over the internet. • Use the device to visit specific Web sites and perform queries, acquire the contents of the device, selectively delete some data, locate and display the URLs of visited sites and any associated data acquired (e.g., images, text, etc.). • Expect that information about most recent Web activity can be found and reported.
Text File Formats	Determine whether the tool can find and display a compilation of text files residing on the device, including deleted files. • Load the device with various types of text files, (via email and device synchronization protocols), selectively delete some files, acquire the contents of the device, find and report the data. • Expect that all files with common text file formats (i.e., .txt, .doc, .pdf) can be found and reported, if not deleted. Expect that remnants of deleted information can be recovered and reported.
Graphics File Formats	Determine whether the tool can find and display a compilation of the graphics formatted files residing on the device, including deleted files. • Load the device with various types of graphics files, (via email and device synchronization protocols) selectively delete some files, acquire the contents of the device, locate and display the images. • Expect that all files with common graphics files formats (i.e., .bmp, .jpg, .gif, .tif, and .png) can be found, reported, and collectively displayed, if not deleted. Expect that remnants of deleted information can be recovered and reported.
Compressed Archive File Formats	Determine whether the tool can find text, images, and other information located within compressed-archive formatted files (i.e., .zip, .rar, .tar, .tgz, and self-extracting .exe) residing on the device. • Load the device with various types of file archives, (via email and device synchronization protocols) acquire the contents of the device, find and display selected filenames and file contents. • Expect that text, images, and other information contained in the compressed archive formatted files can be found and reported.

Scenario	Description
Misnamed Files	Determine whether the tool can recognize file types by header information instead of file extension, and find common text and graphics formatted files that have been misnamed with a misleading extension. • Load the device (via email and device synchronization protocols) with various types of common text (e.g., .txt) and graphics files (e.g., .bmp, .jpg, .gif, and .png) purposely misnamed, acquire the contents of the device, locate and display selected text and images. • Expect that all misnamed text and graphics files residing on the device can be recognized, reported, and, if an image, displayed.
Peripheral Memory Cards	Determine whether the tool can acquire individual files stored on a memory card inserted into the device and whether deleted files would be identifiable and recoverable. • Insert a formatted memory card containing text, graphics, archive, and misnamed files into an appropriate slot on the device, delete some files, acquire the contents of the device, find and display selected files and file contents, including deleted files. • Expect that the files on the memory card, including deleted files, can be properly acquired, found, and reported in the same way as expected with on-device memory.
Acquisition Consistency	Determine if the tool provides consistent hashes on files resident on the device for two back-to-back acquisitions • Acquire the contents of the device and create a hash over the memory, for physical acquisitions, and over individual files, for logical acquisitions. • Expect that hashes over the individual file hashes are consistent between the two acquisitions, but inconsistent for the memory hashes.
Cleared Devices	Determine if the tool can acquire any user information from the device or peripheral memory, after a hard reset has been performed. • Perform a hard reset on the device, acquire its contents, and find and display previously available filenames and file contents. • Expect that no user files, except those contained on a peripheral memory card, if present, can be recovered.
Power Loss	Determine if the tool can acquire any user information from the device, after it has been completely drained of power. • Completely drain the device of power by exhausting the battery or removing the battery overnight and then replacing, acquire device contents, and find and display previously available filenames and file contents. • Expect that no user files, except those contained on a peripheral memory card, if present, can be recovered.

A distinct set of scenarios was developed for SIM forensic tools. The SIM scenarios differ from the phone scenarios in several ways. SIMs are highly standardized devices whose interface, behavior, and content are relatively uniform. All of the SIM tools broadly support any SIM for acquisition via an external reader. Thus, the emphasis in these scenarios is on loading the memory of the SIM with specific kinds of information for recovery, rather than the memory of the handset. Once a scenario is completed using a suitable GSM phone or SIM management program, the SIM can be processed by each of the SIM tools in succession. Table 8 gives an overview of the SIM scenarios, including their purpose, method of execution, and expected results.

Table 8: SIM Scenarios

Scenario	Description
Basic Data	Determine whether the tool can recover subscriber (i.e., IMSI, ICCID, SPN, and LP elementary files), PIM (i.e., ADN elementary file), call (i.e., LND elementary file), and SMS message related information on the SIM, including deleted entries, and whether all of the data is properly decoded and displayed. • Populate the SIM with known PIM, call, and SMS message related information that can be verified after acquisition; then remove the SIM for acquisition and analysis. • Expect that all information residing on the SIM can be successfully acquired and reported.
Location Data	Determine whether the tool can recover location-related information (i.e., LOCI, LOCIGPRS, and FPLMN elementary files), on the SIM, and whether all of the data is properly decoded and displayed. Location information can indicate where the device was last used for a particular service and other networks it might have encountered. • Register location-related data maintained by the network on the SIM by performing voice and data operations at known locations, then remove the SIM for acquisition and analysis. • Expect that all location-related information can be successfully acquired and reported.
EMS Data	Determine whether the tool can recover EMS messages over 160 characters in length and containing non-textual content, and whether all of the data is properly decoded and displayed for both active and deleted messages. EMS messages can convey pictures and sounds, as well as formatted text, as a series of concatenated SMS messages. • Populate the SIM with known EMS content that can be verified after acquisition; then remove the SIM for acquisition and analysis. • Expect that EMS messages can be successfully acquired and reported.
Foreign Language Data	Determine whether the tool can recover SMS messages and PIM data from the SIM that are in a foreign language, and whether all of the data is properly decoded and displayed. • Populate the SIM with known SMS and PIM content that can be verified after acquisition; then remove the SIM for acquisition and analysis. • Expect that EMS and foreign language data can be successfully acquired and reported.

The chapters that follow give a detailed synopsis of each tool and summary of the results of applying the above scenarios to the target devices to determine the extent to which a given tool meets the expectations listed. The tool synopsis concentrates on several core functional areas: acquisition, search, graphics library, and reporting, and also other useful features such as tagging uncovered evidence with a bookmark.

The scenario results for each tool are weighed against the predefined expectations defined above in Table 7 and Table 8, and assigned a ranking. The entry "Meet" indicates that the software met the expectations of the scenario for the device in question. Since the scenarios are acquisition oriented, this ranking generally means that all of the identified data was successfully recovered. One caveat is that some phones lack the capability to handle certain data prescribed under a scenario, in which case the ranking applies only to the relevant subset. Similarly, the entry "Below" indicates that the software fell short of fully meeting expectations, while "Above" indicates that the software surpassed expectations.

A "Below" ranking is often a consequence of a tool performing a logical acquisition and being unable to recover deleted data, which is understandable. However, the ranking may also be due to active data placed on the device not being successfully recovered, which is more of a concern. An "Above" ranking is typically a result associated with the characteristics of a device, such as the reset function not completely deleting data and leaving remnants for recovery by the tool. "Above" rankings should only occur with the last two phone scenarios: Cleared Devices and Power Loss. The entry "Miss" indicates that the software unsuccessfully met any expectations, highlighting a potential area for improvement. Finally, the entry "NA" indicates that a particular scenario was not applicable to the device.

Synopsis of PDA Seizure

PDA Seizure version 3.0.3.89 is able to acquire information from either Pocket PC, Palm OS or BlackBerry devices, including those with cellular capabilities. However, it is not specifically oriented toward cellular phones and omits certain features such as SIM acquisition and examination for GSM phones. PDA Seizure allows the examiner to connect a device via a USB or a Serial connection. Examiners must have the correct cables and cradles to ensure connectivity, compatible synchronization software, and a backup battery source available. Synchronization software (e.g., Microsoft ActiveSync, Palm HotSync, BlackBerry desktop manager software) allows examiners to create a guest partnership between the forensic workstation and the device being investigated.

Pocket PC Phones

The acquisition of a Pocket PC Mobile phone device is done through PDA Seizure with the aid of Microsoft's ActiveSync communication protocol. During the ActiveSync connection an examiner creates a connection as a "Guest" to the device. The "Guest" account is essential for disallowing any synchronization between the PC and the device before acquisition. Before the acquisition of information begins, PDA Seizure places a small dll program file on the device in the first available block of memory, which is then removed at the end of acquisition. Paraben indicated that PDA Seizure uses the dll to access unallocated regions of memory on the device.

To get the remaining information, PDA Seizure utilizes Remote API (RAPI)[9], which provides a set of functions for desktop applications to communicate with and access information on Windows CE platforms. These functions are accessible once a Windows CE device is connected through ActiveSync. RAPI functions are available for the following:

- Device system information – includes version, memory (total, used, and available), and power status retrieval
- File and directory management – allows retrieval of path information, find specific files, permissions, time of creation, etc.
- Property database access – allows information to be gleaned from database information present on the device
- Registry manipulation – allows the registry to be queried (i.e., keys and associated value)

If the device is password protected, the correct password must be supplied before the acquisition stage begins, as illustrated below in Figure 4. If the correct password is not known or provided, connectivity cannot be established and the contents of the device cannot be acquired.

---

[9] Additional information on RAPI can be found at: **http://www.cegadgets.com/artcerapi.htm**

Figure 4: Password Prompt (Pocket PC)

During the beginning stages of acquisition, the examiner is prompted with four choices of data to acquire as illustrated below.

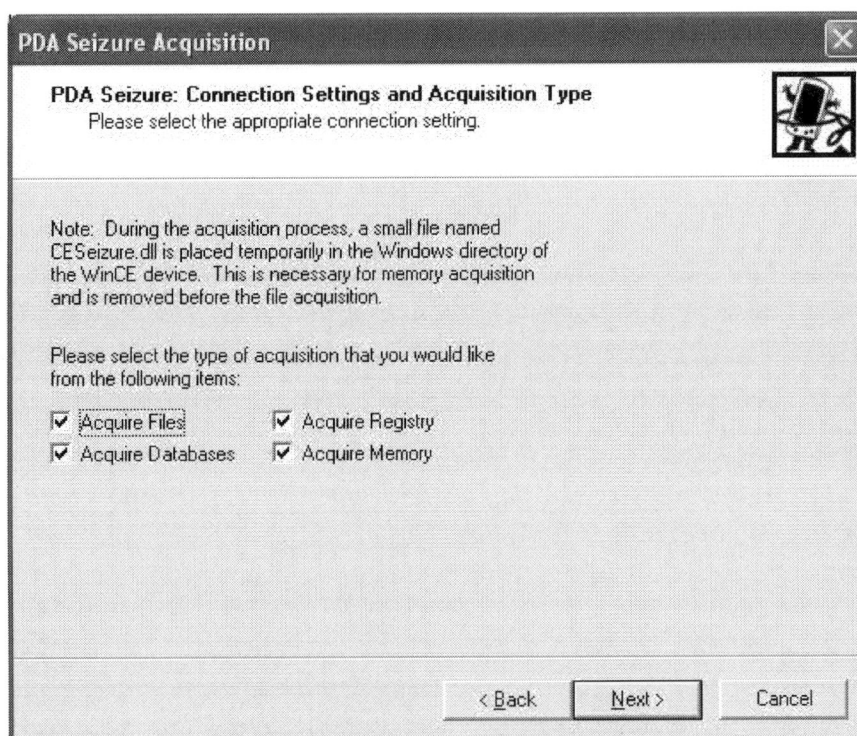

Figure 5: Acquisition Selection (Pocket PC)

Palm OS Phones

The acquisition of a Palm OS device with cell phone capabilities entails the forensic examiner exiting all active HotSync applications and placing the device in console mode. Console mode is used for physical acquisition of the device.[10] To put the Palm OS device in console mode, the examiner must go to the search window (press the magnifying glass by the Graffiti writing area), enter via the Graffiti interface the following symbols: lower-case cursive L, followed by two dots (results in a period), followed by writing a "2" in the number area. For acquiring data from a

---

[10] Additional information on console mode can be found at: **http://www.ee.ryerson.ca/~elf/visor/dot-shortcuts.html**

28

palmOne Treo 600, the technique used is slightly different. Instead of entering console mode via the Graffiti writing area, the shortcut used must be entered via the QWERTY keyboard. Console mode is device specific and the correct sequence of graffiti characters can be found at the manufacturer's Web site. If the device is password protected, the proper password must be entered before acquisition. During the beginning stages of acquisition the examiner is prompted with four choices of data to acquire as illustrated below.

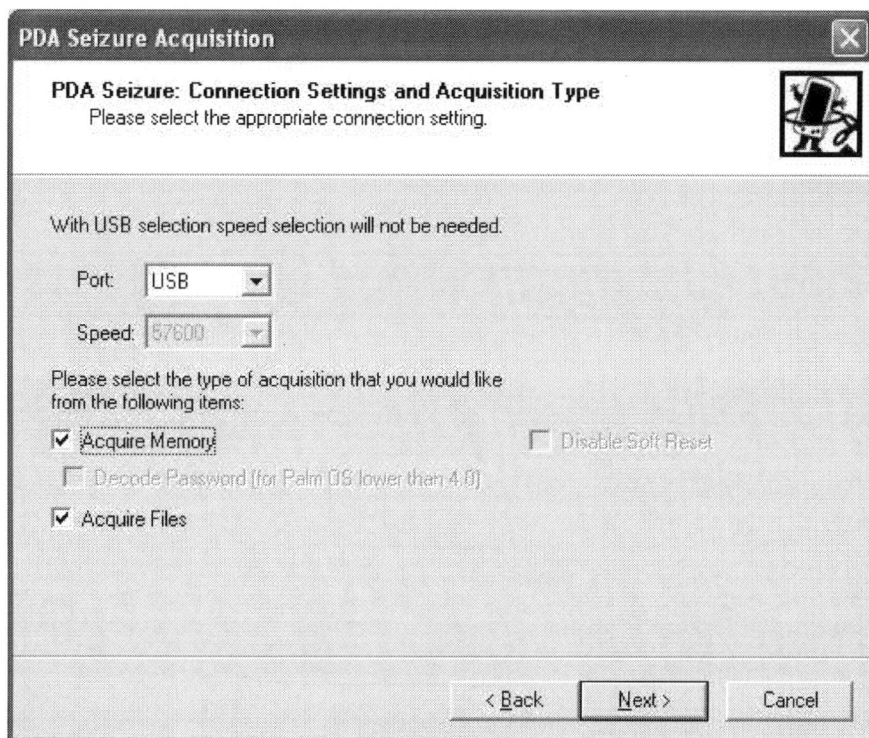

PDA Seizure Acquisition

PDA Seizure: Connection Settings and Acquisition Type
Please select the appropriate connection setting.

With USB selection speed selection will not be needed.

Port: USB

Speed: 57600

Please select the type of acquisition that you would like
from the following items:

☑ Acquire Memory   ☐ Disable Soft Reset

☐ Decode Password (for Palm OS lower than 4.0)

☑ Acquire Files

< Back   Next >   Cancel

Figure 6: Acquisition Selection (Palm OS)

BlackBerry Devices

The acquisition of a BlackBerry device is done through PDA Seizure without the aid of synchronization protocols or the BlackBerry Desktop Manager. The BlackBerry Desktop Manager does allow users to upload applications, perform backups and restorations, and synchronization of defined data (e.g., Address Book, Calendar, Memo Pad, Tasks). Below is a dialog box presented to the examiner before acquisition begins, if the BlackBerry device is password protected. If the password is not correctly supplied within 10 attempts all data is lost from the device and the BlackBerry OS has to be reinstalled.

Figure 7: Password Prompt (BlackBerry)

After device selection the examiner is prompted with the following options of acquiring either individual databases, memory, or both.

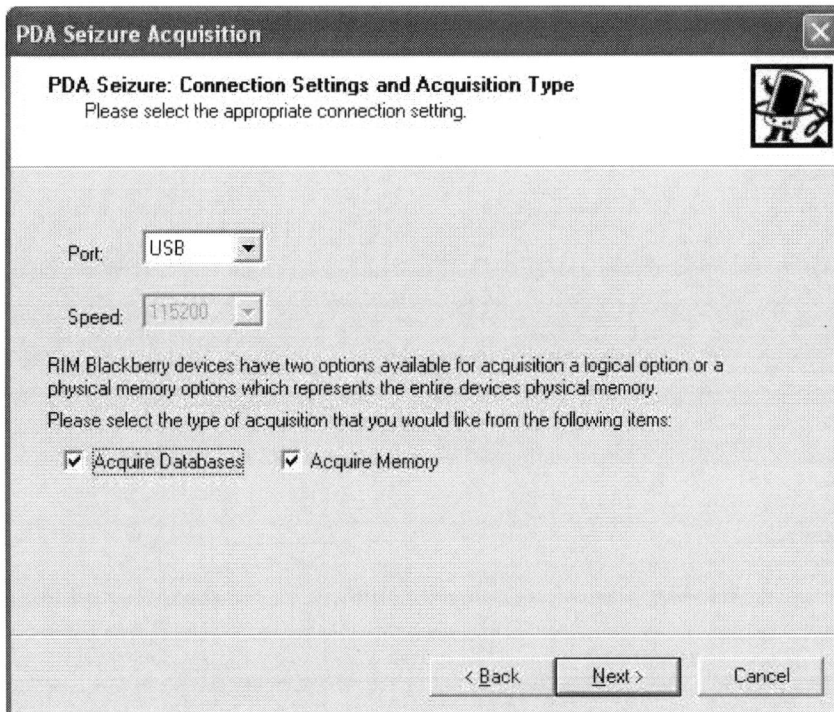

Figure 8: Acquisition Selection (BlackBerry)

If Acquire Databases and Memory are both selected, the memory is acquired first and then the examiner is alerted that a soft reset will be performed as illustrated below before individual databases are acquired. The soft reset does not affect the device integrity or the integrity of the data associated with the acquisition.

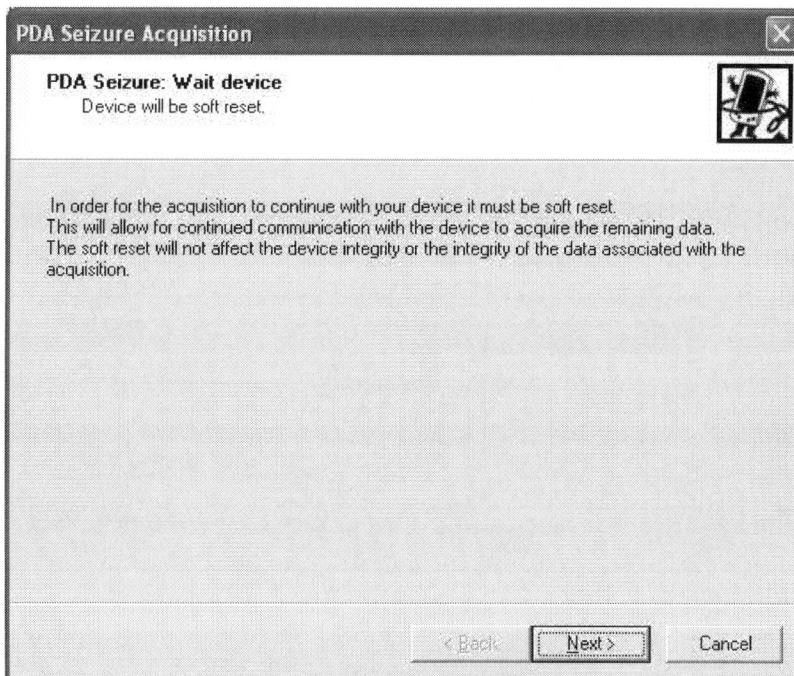

Figure 9: Soft Reset (BlackBerry)

Acquisition Stage

Two methods exist to begin the acquisition of data from the PDA device. The acquisition can be enacted through the toolbar using the Acquire icon or through the Tools menu and selecting Acquire Image. Either option starts the acquisition process. With the acquisition process, both files and memory images can be acquired. By default, the tool marks both types of data to be acquired. Once the acquisition process is selected, the acquisition wizard illustrated below in Figure 10 guides the examiner through the process.

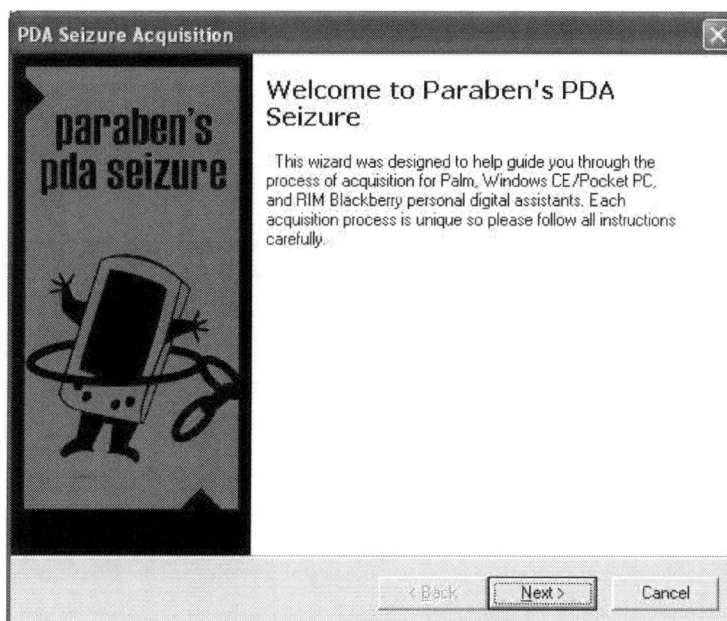

Figure 10: Acquisition Wizard

After clicking next on the Acquisition Wizard, the examiner is prompted to select which type of device to acquire, as illustrated below.

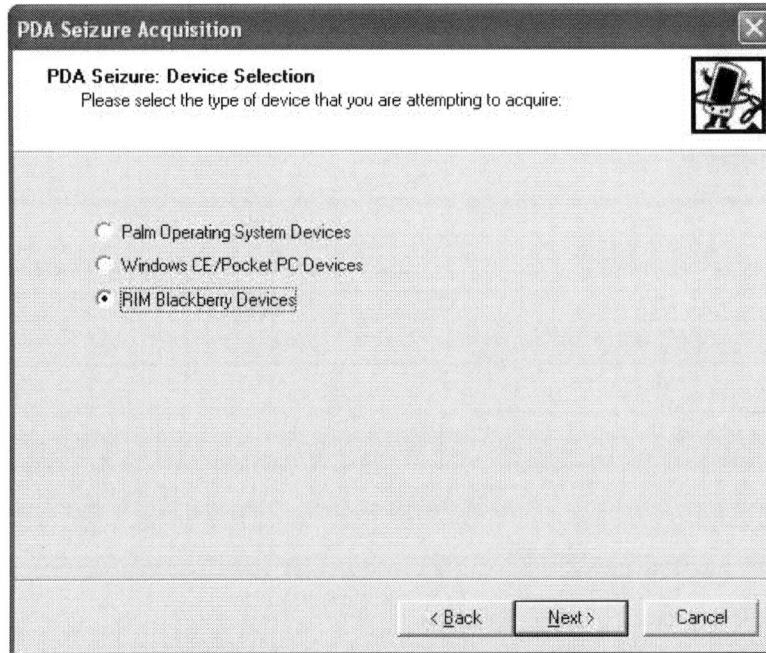

Figure 11: Acquisition Selection

Figure 12 below contains an example screen shot of PDA Seizure during the acquisition of a Pocket PC device, displaying the various fields provided by the interface.

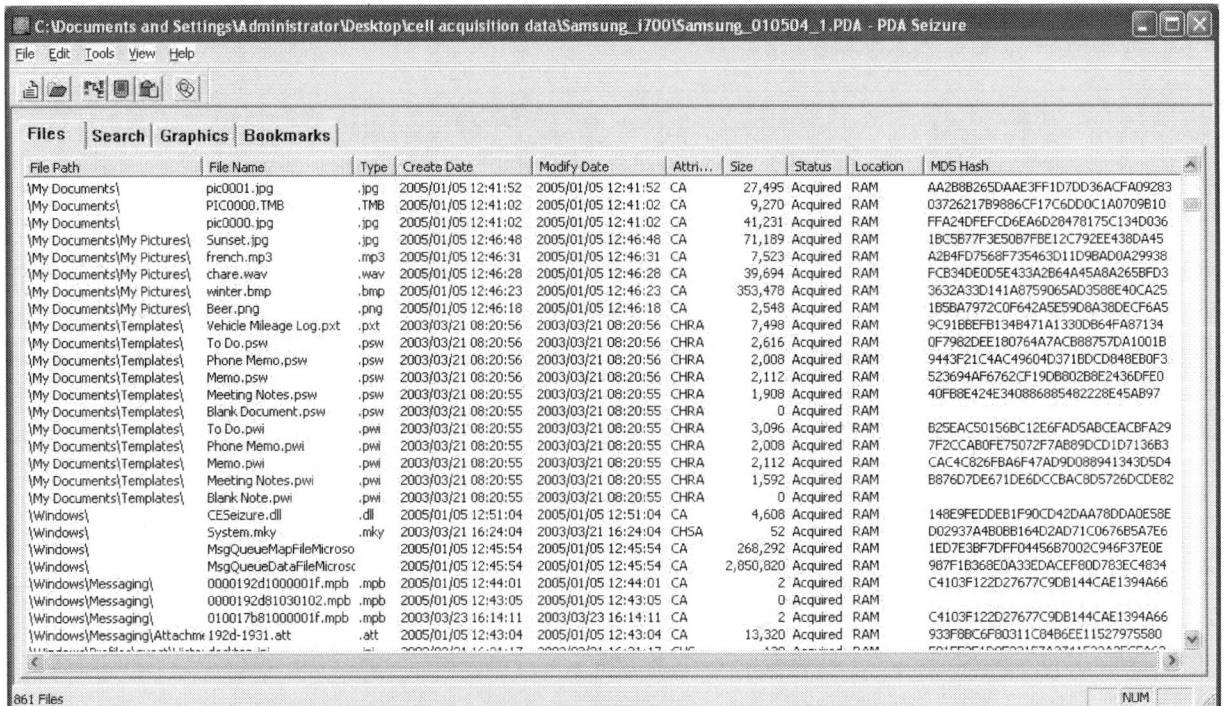

Figure 12: Acquisition Screen Shot (Pocket PC)

After Pocket PC acquisition, PDA Seizure reports the following for each individual files: File Path, File Name, File Type, Creation and Modification Dates, File Attributes, File Size, Status, and an MD5 File Hash. Validation of file hashes taken before and after acquisition can be used to detect whether files have been modified during the acquisition stage.

During the acquisition process, CESeizure.dll allows unallocated memory regions to be acquired. The examiner is prompted with check boxes to select one or all of the following before acquiring information on the Pocket PC device: Acquire Files, Acquire Databases, Acquire Registry, and/or Acquire Memory. Each file acquired can be viewed in either text or hex mode, allowing examiners to inspect the contents of all files present. Examiners must use one of the following options to view the files: export the file, launch a Windows application corresponding to the file extension (Run File's Application), or, for Palm OS devices, view the file through the POSE.

Search Functionality

PDA Seizure's search facility allows examiners to query files for content. The search function searches the content of files and reports all instances of a given string found. The screen shot shown below in Figure 13 illustrates an example of the results produced for the string "homer simpson."

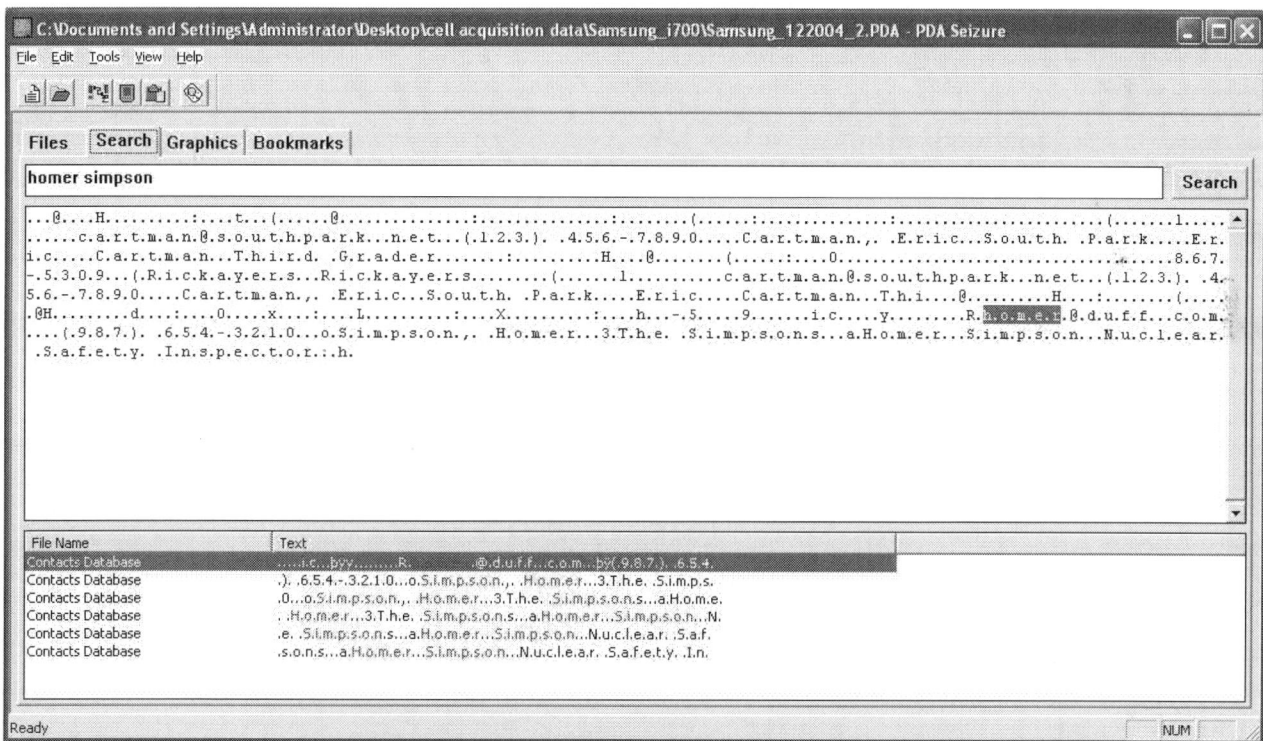

Figure 13: File Content String Search (Pocket PC)

Additionally the search window provides an output of memory related to the string search provided by the examiner. This allows examiners to scroll through sections of memory and bookmark valuable information for reporting to be used in judicial, disciplinary, or other proceedings.

Graphics Library

The graphics library enables examiners to examine the collection of graphics files present on the device, identified by file extension. Deleted graphics files do not appear in the library. Manually performing identification of graphics files based on file signature, a more precise approach, is very time-consuming and may cause vital data to be omitted. If deleted graphics files exist, they must be identified via the memory window by performing a string search to identify file remnants. However, recovery of the entire image is difficult, since its contents may be compressed by the filesystem or may not reside in contiguous memory locations, and some parts may be unrecoverable. It also requires knowledge of associated data structures to piece the parts together successfully. Figure 14 shows a screen shot of images acquired from a Samsung i700 device.

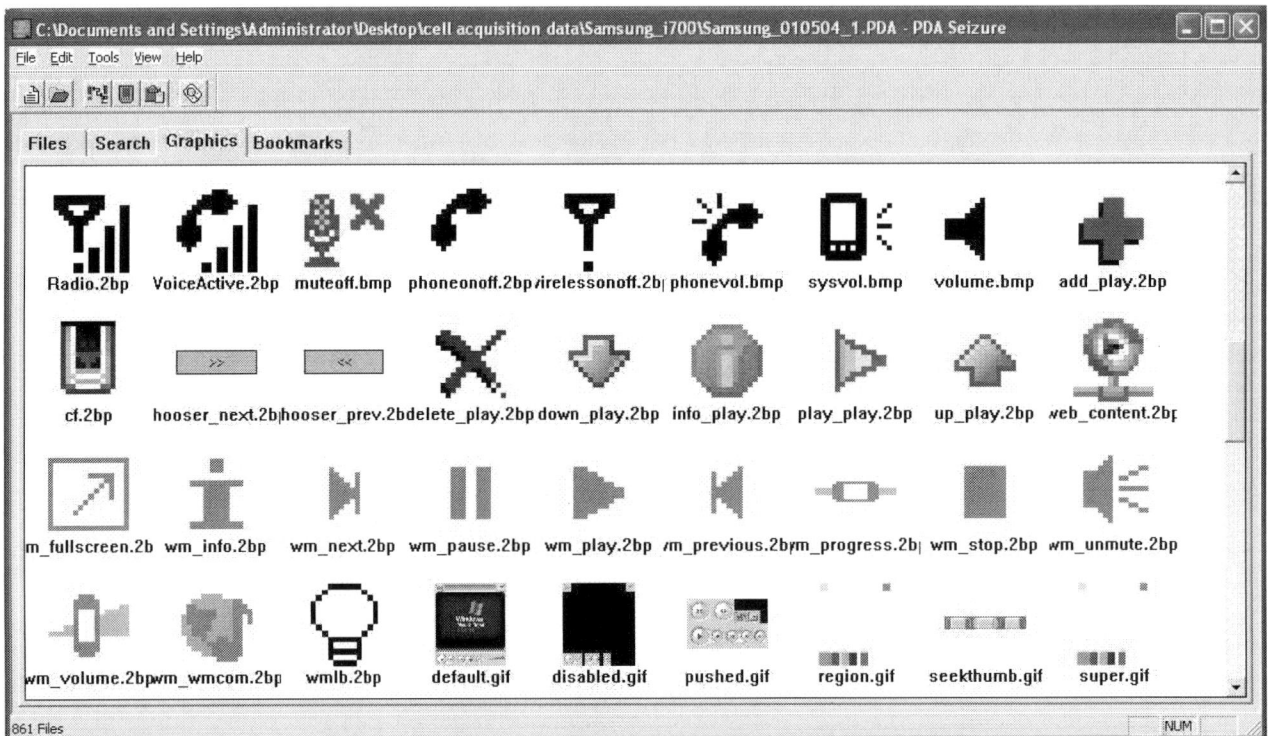

Figure 14: Graphics Library (Pocket PC)

Bookmarking

During an investigation, forensic examiners often have an idea of the type of information for which they are looking, based upon the circumstances of the incident and information already obtained. Bookmarking allows forensic examiners to mark items that are relevant to the investigation. Such a capability gives the examiner the means to generate case specific reports containing significant information found during the examination, in a format suitable for presentation. Bookmarks can be added for multiple pieces of information found and each individual file can be exported for further analysis if necessary. Illustrated in Figure 15 below is an example of various bookmarks created after acquisition. As mentioned earlier, the files found and bookmarked can be exported to the PC and viewed with an application suitable for the type of file in question.

34

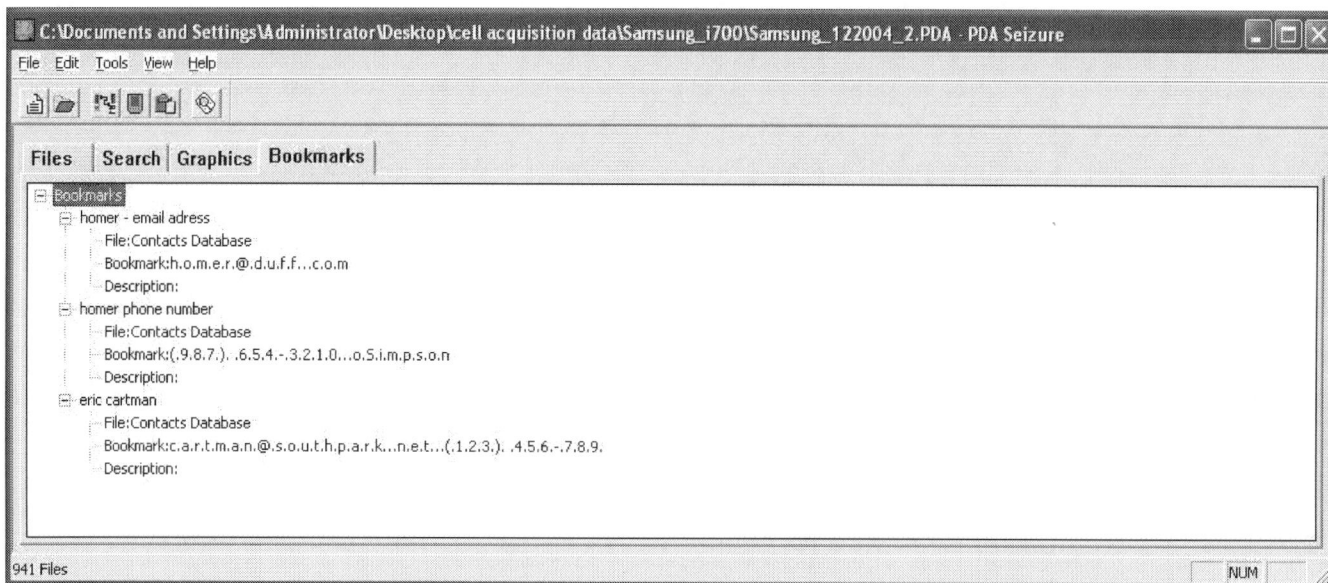

Figure 15: Bookmark Creation (Pocket PC)

Additional Tools

Export All Files: Examiners can export all files reported after the acquisition stage has been completed. After the files have been exported, a folder is created, based upon the case file name, with two subfolders: one each for RAM and ROM. Depending upon the type of file, the contents can be viewed with an associated desktop application or with a device specific emulator.

Validate Hash Codes: The Validate Hash Codes option can be run after a successful acquisition and is designed to report files that have been altered during the acquisition process.

Figure 16: Validate Hash Codes (Pocket PC)

PDA File Compare: PDA Seizure has a built-in function that compares acquisition files. To operate the compare feature, one file is loaded into the program then compared via the Tools menu option to another file. The files are compared based on hash codes. The results are shown in a dialog box listing the file name, the result of the compare, and the size in each .pda file. Double-clicking a file, or highlighting a file and clicking the "Show Files" button, pops up a side-by-side hex view of the two files with the differences shown in red. PDA File Compare is illustrated below in Figure 17.

35

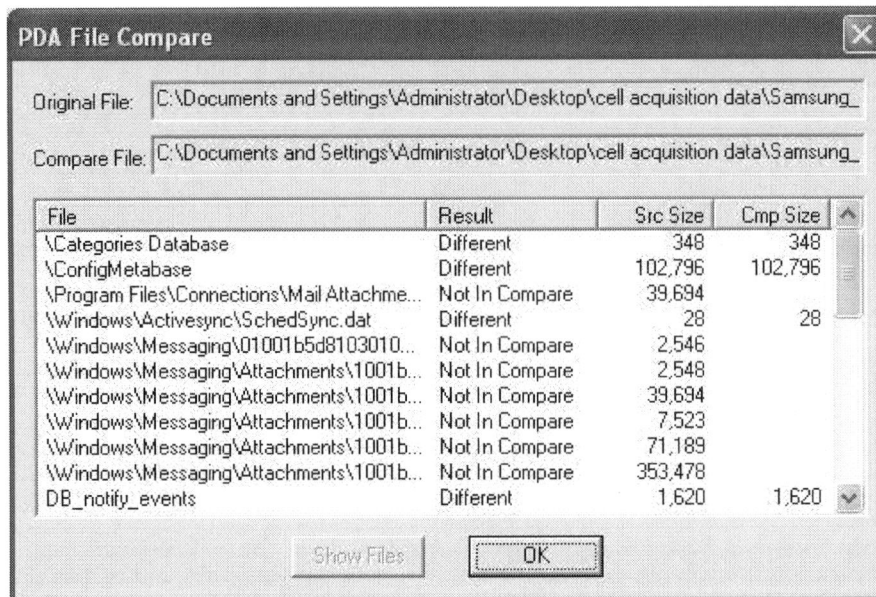

Figure 17: PDA File Compare (Palm OS)

PDA Seizure File Viewer: Files that have not been deleted have the option of being viewed in either text or hex, or with the "Run File's Application" function, which calls an associated application to display the data on the examiner's local machine. The latter allows graphics and other file types that are not in a standard flat ASCII file format to be viewed.

Report Generation

Reporting is an important task for examiners. PDA Seizure provides a user interface for report generation that allows examiners to enter and organize case-specific information. Each case contains an identification number and other information specific to the investigation for reporting purposes, as illustrated in Figure 18 below.

Once the report has been generated, it produces an .html file for the examiner, including bookmarked files, total files acquired, acquisition time, device information, etc. If files were modified during the acquisition stage, the report identifies them.

36

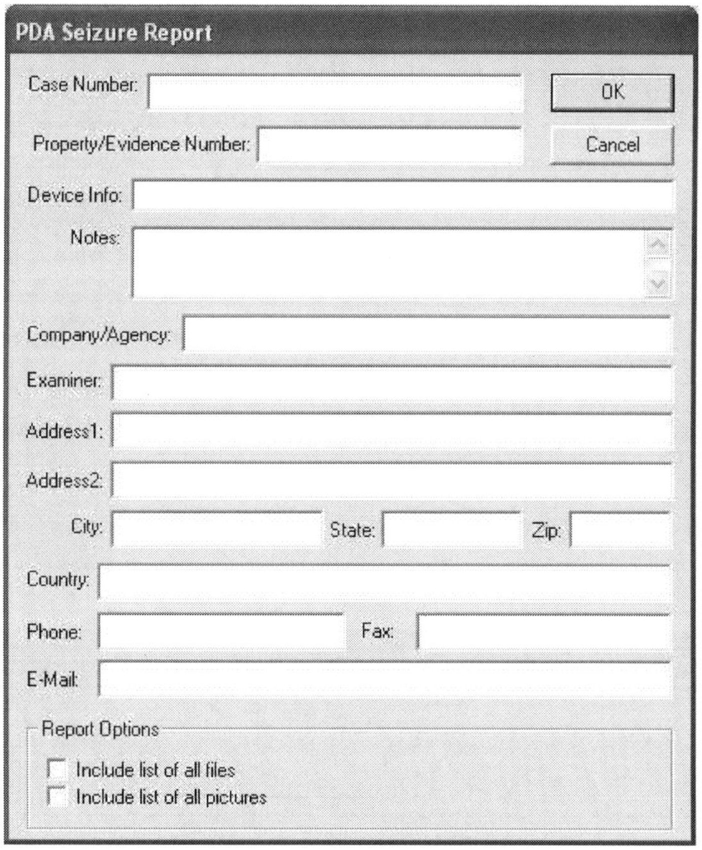

Figure 18: Report Generation

Scenario Results

Table 9 summarizes the results from applying the scenarios listed at the left of the table to the devices across the top. More information can be found in Appendix A: PDA Seizure Results.

Table 9: Results Matrix

Scenario	Device					
	BlackBerry 7750	BlackBerry 7780	Kyocera 7135	Motorola MPX220	Samsung i700	Treo 600
Connectivity and Retrieval	Meet	Meet	Meet	Meet	Meet	Meet
PIM Applications	Meet	Meet	Meet	Miss	Meet	Meet
Dialed/Received Phone Calls	Meet	Meet	Meet	Miss	Meet	Meet
SMS/MMS Messaging	Meet	Meet	Below	Below	Below	Meet
Internet Messaging	Meet	Meet	Below	Below	Below	Below
Web Applications	NA	Below	Below	Below	Below	Below
Text File Formats	Below	Below	Meet	Below	Below	Meet
Graphics File Formats	Miss	Miss	Miss	Below	Below	Miss
Compressed Archive File Formats	Miss	Miss	Miss	Meet	Meet	Below
Misnamed Files	Miss	Miss	Meet	Meet	Meet	Meet

Scenario	Device					
	BlackBerry 7750	BlackBerry 7780	Kyocera 7135	Motorola MPX220	Samsung i700	Treo 600
Peripheral Memory Cards	NA	NA	Miss	Below	Below	Miss
Acquisition Consistency	Meet	Meet	Below	Meet	Below	Below
Cleared Devices	Meet/Above	Meet/Above	Meet	Meet	Meet	Meet
Power Loss	Above	Above	Meet	Above	Meet	Meet

Synopsis of Pilot-link

The pilot-link software can be used to obtain both the ROM and RAM from Palm OS devices. The data can be imported into the Palm OS Emulator (POSE), allowing a virtual view of the data contained on the device, or the individual files can be viewed with a standard ASCII, hex editor or through a compatible forensic application. Additionally, the data created from pilot-link can be imported into other compatible forensic applications. Once the software is installed and configured, communications between the PC and the device can begin. RAM and ROM are dumped from the device with the following commands: pi-getrom and pi-getram.

In order to prepare data to be imported into POSE the following commands are issued:
- pi-getrom: Generates a ROM image of the device.
- pi-getram: Generates a RAM image of the device.
- pilot-xfer –b <dir>: Backs up databases (i.e., .prc, .pdb, and .pqa) to a file, which can be imported elsewhere.

A few other useful pilot-link commands are the following:
- addresses: Dumps the Palm OS address book.
- memos: Exports memos from Palm OS in a standard mailbox format.
- pilot-clip: Exports data from the Palm OS clipboard.
- pilot-file: Dissects and allows a view of detailed information about the Palm Resource Database, Palm Record Database and the Palm Query Application files.
- pilot-undelete: Turn archived records into normal records.
- pilot-xfer: Backup, restore, install, and delete Palm OS databases.
- read-expenses: Export Palm Pro Expense database into text format.
- read-ical: Export Palm OS Datebook and ToDo databases into an Ican calendar.
- read-todos: Export Palm OS ToDo database into generic text format.
- reminders: Export Palm OS Datebook into a 'remind' data file.

Supported Phones

Pilot-link communicates with Palm OS handheld devices, such as those manufactured by Palm, Handspring, Handera, TRGPro, and Sony, over a serial connection. Palm OS devices that double as a cell phone do not necessarily work properly with all of the integrated functions provided by Pilot-link. For instance, Pilot-link was only able to acquire individual database files from the PalmOne Treo 600. However, Pilot-link could acquire all data from a Kyocera 7135 Palm OS-based cell phone using pi-getram, pi-getrom, and pilot-xfer -b. Before performing an actual acquisition, the examiner should experiment with Palm OS-based phones to determine whether Pilot-link can acquire the necessary data.

Acquisition Stage

To capture all of the data present on a Palm OS compatible device, examiners typically issue the following three Pilot-link commands: pi-getram, pi-getrom, and pilot-xfer –b, which respectively retrieve the contents of RAM and ROM from the device and acquire the contents of the device logically (i.e., individual databases). The acquisition process begins with creating a connection between the forensic workstation and the Palm OS-based cell phone, which requires the proper serial cable. Once a successful connection has been made, Pilot-link uses the HotSync protocol to acquire data from the device. The memory size of the device determines the time involved for acquiring data; typically the acquisition time is lengthy using Pilot-link, since data is transferred over a serial connection. Maintaining power to the device throughout the acquisition process is advised to eliminate the chance of losing data due to battery depletion.

Search Functionality

Pilot-link does not have a built-in search engine, therefore, third-party search facilities must be used. The output data can be queried in several ways, such as using a search tool (e.g., grep), using a standard editor (e.g., WinHex), or importing the data into a forensic tool that provides search facilities (e.g., EnCase).

Graphics Library

Pilot-link does not have built-in support for a graphics library. Acquired data can either be imported into POSE or an additional forensic tool (e.g., EnCase) that can parse files correctly and generate a logical view of the data. Graphic files can be manually identified by file header and displayed using a third-party graphic viewer (e.g., gimp, gqview).

Report Generation

Reporting facilities are not supported internally by Pilot-link. Therefore examiners using Pilot-link as the sole tool for acquiring data must use third-party tools or additional techniques to generate a report of their findings.

Scenario Results

Table 10 summarizes the results from applying the scenarios listed at the left of the table to the devices across the top. More information can be found in Appendix B: Pilot-Link Results.

Table 10: Results Matrix

Scenario	Device	
	Kyocera 7135	Treo 600
Connectivity and Retrieval	Meet	Below
PIM Applications	Meet	Meet
Dialed/Received Phone Calls	Meet	Below
SMS/MMS Messaging	Meet	Below
Internet Messaging	Meet	Below
Web Applications	Below	Below
Text File Formats	Meet	Meet
Graphics File Formats	Below	Below
Compressed Archive File Formats	Miss	Miss
Misnamed Files	Meet	Meet
Peripheral Memory Cards	Miss	Miss
Acquisition Consistency	Below	Below

Scenario	Device	
	Kyocera 7135	Treo 600
Cleared Devices	Meet	Meet
Power Loss	Meet	Meet

Synopsis of Cell Seizure

Cell Seizure version 2.0.0.33660 is able to acquire information from various manufacturers' cell phones, including Motorola, Nokia, Siemens, Siemens, and Sony Ericsson.[11] The specific phone manufacturer and model determine the type of cable connection used (i.e., USB vs. RS-232). Cell Seizure also allows independent acquisition of SIM cards with the included SIM card reader that comes with the purchase of Cell Seizure's Toolbox. The Cell Seizure Toolbox provides all the necessary cables to create connectivity between supported phone models and the forensic workstation.

Supported Phones

The make, model, and type of phone determine how much data, if any, that Cell Seizure can acquire. The version of Cell Seizure used can acquire only Phone Book entries from TDMA-based phones. Cell Seizure can acquire the following information for other phones: Phone Calls Made, Phone Calls Received, Text Messages, and Phonebooks for all supported models. Additionally, Cell Seizure can acquire the following on some supported models: To-Do Lists, Calendar, Call Times, Call Info, Call History, Contact info, Phone Number of Cell, Image Gallery (e.g., Wallpapers, Camera Phone Images, etc.), Ring Tones, Reminders, Memos, Voice Memos, Events, Profiles, Games, Logos, WAP Settings, WAP Bookmarks, GRPS Access Points, Java files, and a complete Memory Dump. The breadth and depth of information acquired depends on the make, model, and network on which the phone operates.

Acquisition Stage

Two methods are provided to begin the acquisition of data from cell phones. The acquisition can be enacted through the toolbar using the Acquire icon or through the Tools menu and selecting Acquisition. Either option starts the acquisition process. The initial stages of acquiring data from a cell phone are guided by an internal wizard, which solicits the following information:

- Manufacturer device selection or GSM SIM card
- Model selection
- Connection type (e.g., USB or serial)
- Data to be acquired (e.g., SMS history, calendar, phone calls, phonebook)

With the acquisition process, both internal phone memory and basic SIM card information (e.g., phone book entries, SMS messages) are acquired. To obtain a more detailed analysis of the data contained on the SIM, the external SIM card reader can be used. Once the acquisition process is selected, the acquisition wizard illustrated below in Figure 19 guides the examiner through the process.

---

[11] Additional information on supported phone models can be found at: **http://www.paraben-forensics.com/cell_models.html**

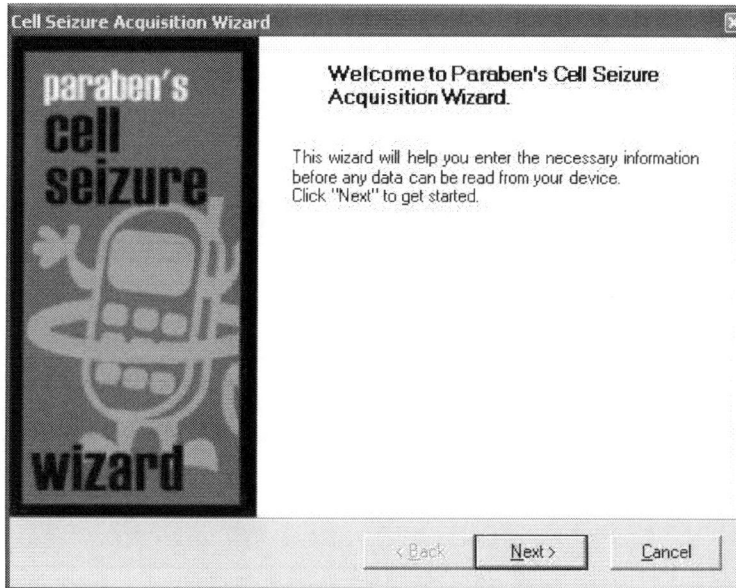

Figure 19: Acquisition Wizard

The next step in the data acquisition process entails selecting either GSM SIM card or the appropriate phone manufacturer entry, as illustrated in Figure 20 below. The former pertains to acquisition of SIM data via an external reader.

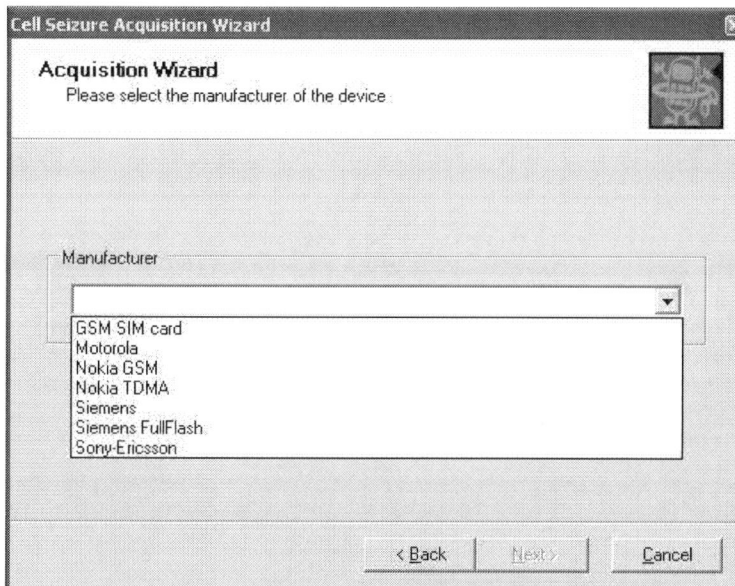

Figure 20: Acquisition Wizard (Manufacturer Selection)

After the proper manufacturer is selected, the user is prompted to select the Connection type (i.e., Serial or USB). The manufacturer and the phone model determine the type of data that can be acquired from the phone. Figure 21 shows the interface displayed for a Nokia GSM phone and the fields that can be selected for capture. To ensure optimal data recovery, Paraben suggests allowing the phone to be powered on for one to three minutes before continuing the acquisition. This allows the SIM to initialize properly. Cell Seizure can also acquire data from certain phones without the SIM being present.

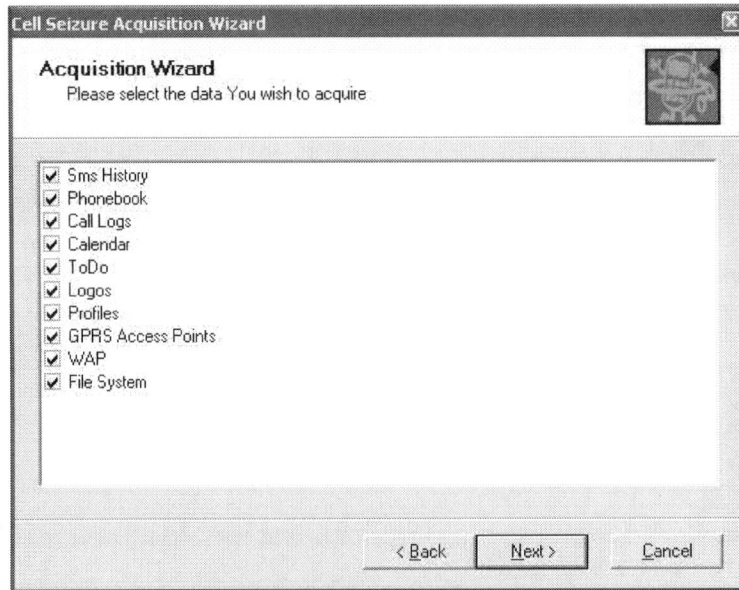

Figure 21: Acquisition Wizard (Nokia GSM)

Search Functionality

Cell Seizure's search facility allows examiners to query the acquired data for content. The search function searches the content of files and reports all instances of a given string found. The screen shot shown in Figure 22 illustrates an example of the search window options and results produced for the string "homer simpson".

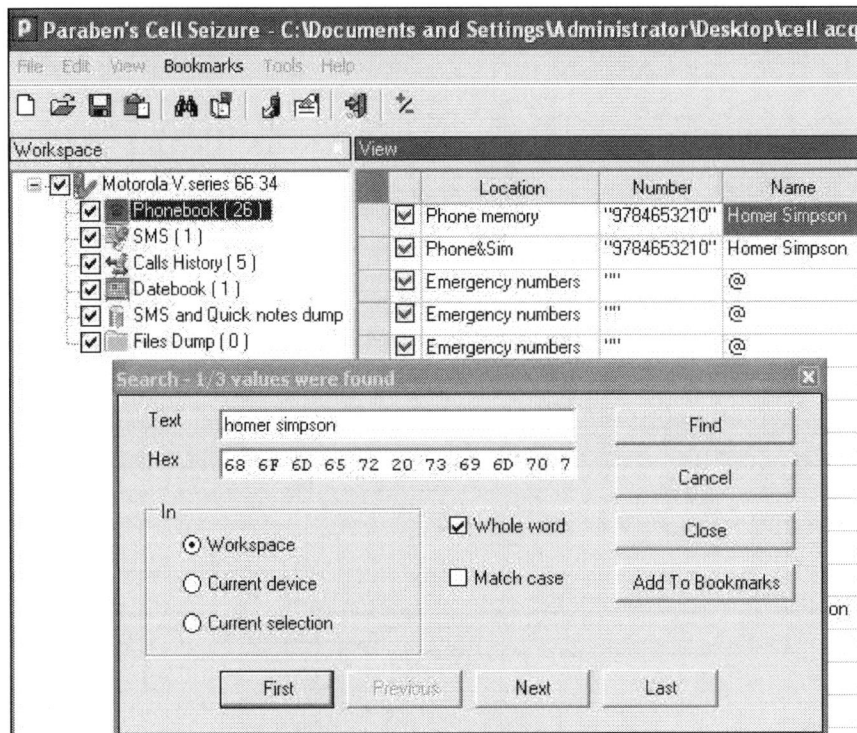

Figure 22: File Content String Search

Graphics Library

The graphics library enables examiners to examine the collection of graphics files present on the device. Each image present can be viewed internally with the Picture Window application allowing examiners to enlarge images if necessary. Additionally, images collected can be exported and inspected with a third party tool, if necessary. Figure 23 shows a screen shot of images acquired from a Nokia 6610i.

Figure 23: Graphics Library

Bookmarking

As mentioned earlier, bookmarking allows forensic examiners to mark items for reporting that are considered relevant to the investigation. Bookmarks can be added for multiple pieces of information found and each individual file can be exported for further analysis if necessary. Illustrated in Figure 24 below is an example of various bookmarks created after acquisition.

45

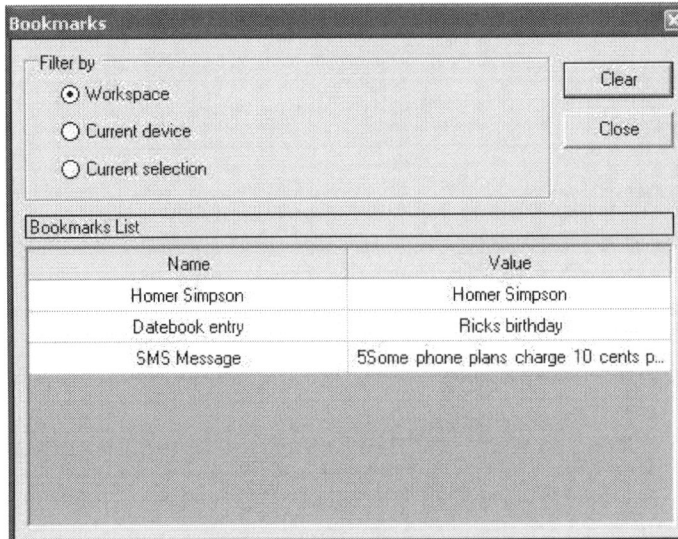

Figure 24: Bookmark Creation

Report Generation

Cell Seizure provides a user interface/report wizard for report generation that allows examiners to enter and organize case specific information. After the device has been acquired, details of the acquisition can be examined by clicking the View Acquisition Report link to launch the report wizard shown in Figure 25. Each report incorporates any information identified by the examiner as pertaining to the investigation (i.e., bookmarked).

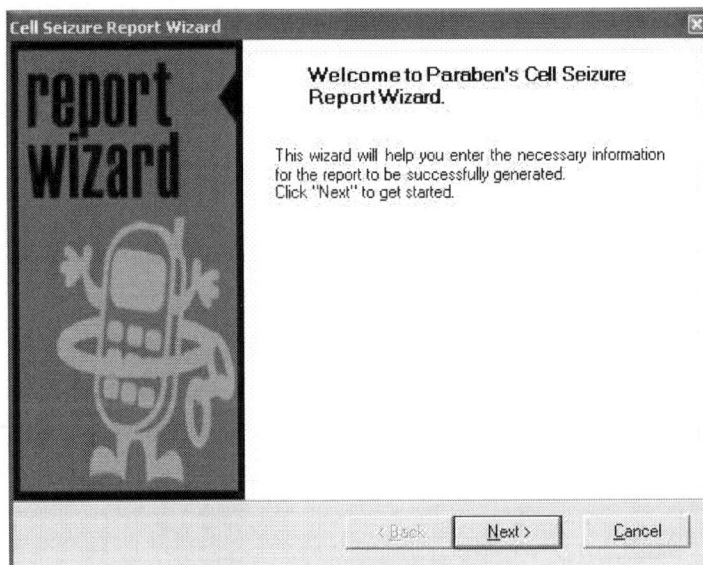

Figure 25: Report Wizard

After the report wizard has been initialized, the examiner has the option of choosing what type of report is generated (i.e., HTML or Text, as shown in Figure 26), and which items to include (i.e., entire workspace vs. selected items only).

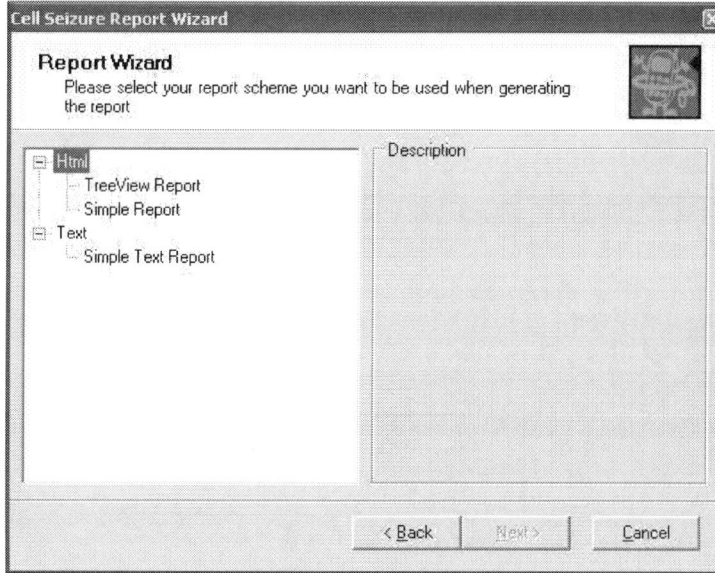

Figure 26: Report Type

HTML generated reports provide two output options: TreeViewReport that generates a hierarchical HTML report with navigation, and a Simple Report that generates a simple HTML report. Graphic files are exported out into a separate folder and the generated HTML file has a hyperlink that allows viewing of individual images. Text-based reports, such as the excerpt in Figure 27, outline examiner-specified data in a text file that can be read with a text editor. All of the reports contain consistent textual-based information, just in different formats.

```
Paraben's Cell Seizure Exported Workspace
        Motorola V.series 66

        Properties
Name------Value------
Manufacturer : Motorola :
Model : V.series 66 :
Serial number :  IMEI449276812531841.... :
MD5 : 47bae0d246ffa06a341b3c62ef61c7ff :
        Phonebook
Location------Number------Name------
Phone memory : "9784653210" : Homer Simpson :
Phone&Sim : "9784653210" : Homer Simpson :
        Properties
Name------Value------
MD5 : d41d8cd98f00b204e9800998ecf8427e :
        SMS
Number------Status------Date/Time------Message------
 "2404016148" :  "STO UNSENT" :  : 体 :
        Properties
Name------Value------
MD5 : d41d8cd98f00b204e9800998ecf8427e :
        Calls History
Name------Number------Direction------
Homer Simpsons/W : "9874653210" : Dialled calls :
 : "301975XXXX" : Dialled calls :
 : "301975XXXX" : Dialled calls :
        Properties
Name------Value------
MD5 : d41d8cd98f00b204e9800998ecf8427e :
        Datebook
Ricks birthday : 0 : 0 : 2000-01-30 00:00 : 1440 :  : non reoccurring :
        Properties
Bookmarks     Homer Simpson : Homer Simpson     Datebook entry : Ricks birthday
```

Figure 27: Text-Based Report Excerpt

Scenario Results

Table 11 summarizes the results from applying the scenarios listed at the left of the table to the devices across the top. More information can be found in Appendix C: Cell Seizure Results.

Table 11: Results Matrix

Scenario	Device				
	Ericsson T68i	Motorola C333	Motorola V66	Nokia 3390	Nokia 6610i
Connectivity and Retrieval	Meet	Meet	Meet	Meet	Meet
PIM Applications	Below	Meet	Below	Below	Below
Dialed/Received Phone Calls	Below	Below	Below	Below	Below
SMS/MMS Messaging	Below	Meet	Meet	Below	Below
Internet Messaging	Miss	NA	NA	NA	NA
Web Applications	Miss	NA	NA	Miss	Below
Text File Formats	NA	NA	NA	NA	Below
Graphics File Formats	Miss	NA	NA	NA	Below

Scenario	Device				
	Ericsson T68i	Motorola C333	Motorola V66	Nokia 3390	Nokia 6610i
Compressed Archive File Formats	NA	NA	NA	NA	Meet
Misnamed Files	NA	NA	NA	NA	Meet
Peripheral Memory Cards	NA	NA	NA	NA	NA
Acquisition Consistency	Meet	Meet	Below	Meet	Meet
Cleared Devices	Above	Above	Meet	NA	NA
Power Loss	Above	Above	Above	Above	Above

SIM Card Acquisition

Paraben's Cell Seizure allows examiners the ability to acquire data directly from a SIM card with the use of the Cell Seizure RS-232 SIM Card Reader. The acquisition steps followed to acquire data directly from the SIM are the same as acquiring data from a phone except for selecting GSM Sim Card, illustrated earlier in Figure 20. The data fields acquired (e.g., Abbreviated Dialing Numbers, Fixed Dialing Numbers, Last Numbers Dialed, SIM Service Dialing Numbers, Short Messages, etc.) depend on the SIM and service provider. The Search Functionality, Bookmarking facilities, and Report Generation operate on the acquired data in a similar fashion to phone acquisitions, described above.

Table 12 summarizes the results from applying the scenarios listed at the left of the table to the SIMs across the top. More information can be found in Appendix N: Cell Seizure – External SIM Results.

Table 12: SIM Card Results Matrix - External Reader

Scenario	SIM		
	5343	8778	1144
Basic Data	Meet	Miss	Meet
Location Data	Meet	Miss	Meet
EMS Data	Below	Miss	Below
Foreign Language Data	Below	Miss	Below

Synopsis of GSM .XRY

Micro Systemation's GSM .XRY version 2.1 (currently up to version 2.5) can acquire information from various manufacturers of GSM cell phones (i.e., Ericsson, Motorola, Nokia, Siemens).[12] The .XRY unit provides the ability to perform acquisitions via cable, IrDA (Infrared), or Bluetooth interfaces. The GSM .XRY unit provides all the necessary cables to create connectivity between supported phone models and the forensic workstation.

Supported Phones

The make, model, and type of phone determine what data GSM .XRY can acquire. Micro Systemation's Web site provides a link to a soft copy version of the manual listing the supported make and models of cell phones. GSM .XRY is targeted at GSM, CDMA and 3G devices. TDMA phones are not currently supported. GSM .XRY can acquire the following information from GSM (Global System for Mobile Communications) phones: Contacts, Calls, Calendar, SMS, Pictures, Audio, Files, Notes, MMS, Video, Network Information and Tasks. Each type of data acquired appears in the menu bar.

Acquisition Stage

Two methods exist to acquire data from cell phones. The acquisition can be enacted through the toolbar, using the Extract data icon, or through the File menu, selecting Extract data from the device. Either option starts the acquisition process. With the acquisition process, both internal phone memory and basic SIM card information (e.g., phone book entries, SMS messages) are acquired. Once the acquisition process is selected, the acquisition wizard illustrated below in Figure 28 appears to guide the examiner through the process.

Figure 28: Acquisition Wizard

---

[12] Additional information on supported phone models can be found at: **www.msab.com**

Following the execution of the acquisition wizard, the examiner selects the acquisition type: mobile device or SIM/USIM card. Assuming selection of a mobile phone device, the examiner is presented with three interface choices to create a connection with the cell phone. Micro Systemation provides specific recommendations (i.e., Cable, Infrared, or Bluetooth) for each make and model of phones supported.

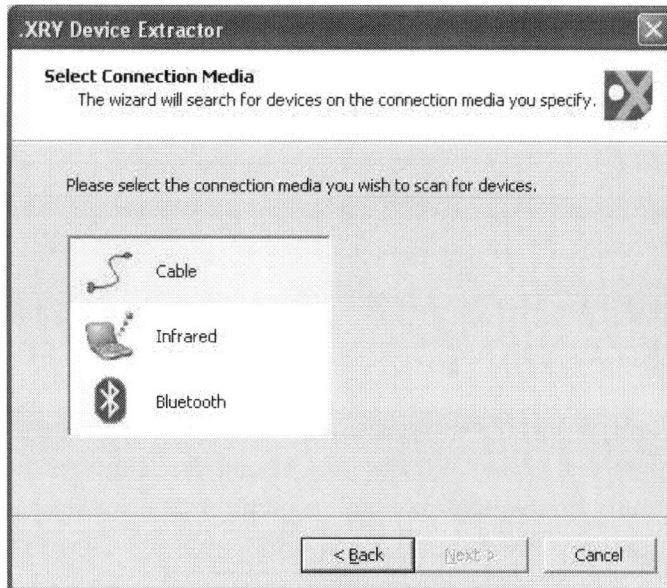

Figure 29: Interface Selection

After a successful connection has been established, the device is identified and the actual acquisition begins.

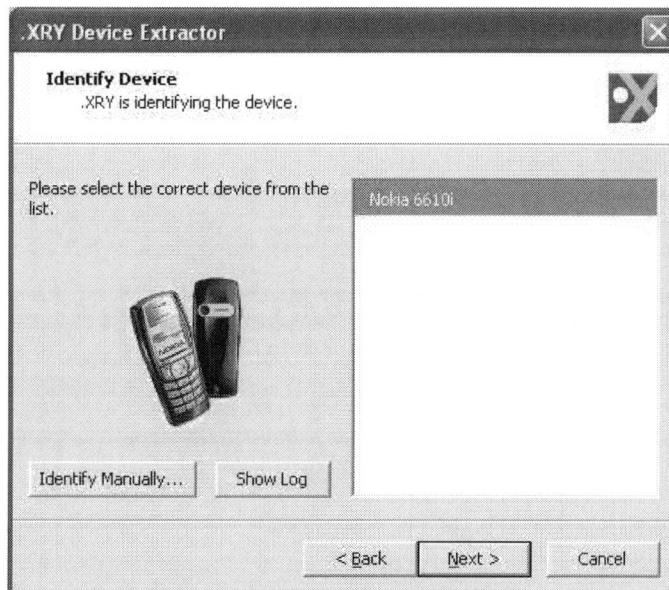

Figure 30: Device Identification

During the acquisition stage GSM .XRY keeps a process log of the status of information extracted from the device, as illustrated in Figure 31.

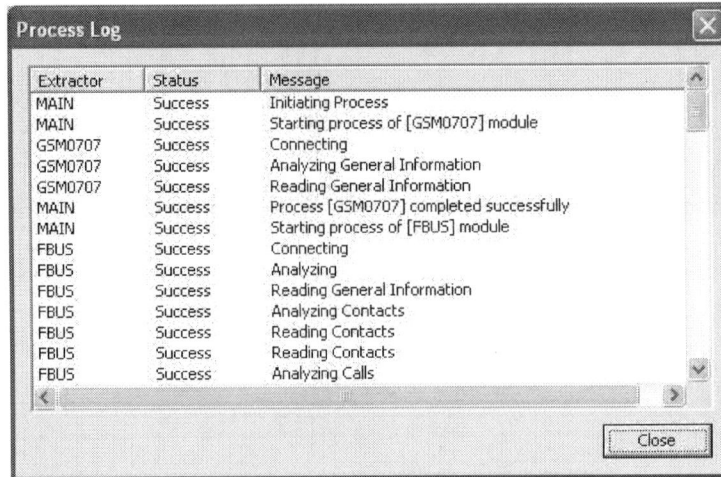

Figure 31: Acquisition Log

Search Functionality

GSM .XRY's search facility allows examiners to query the acquired data for content. The search function searches the content of files and reports all instances of a given string found. The screen shot shown in Figure 32 illustrates an example of the search window options and results produced for the string "SMS message". Search hits that are found are highlighted in pink.

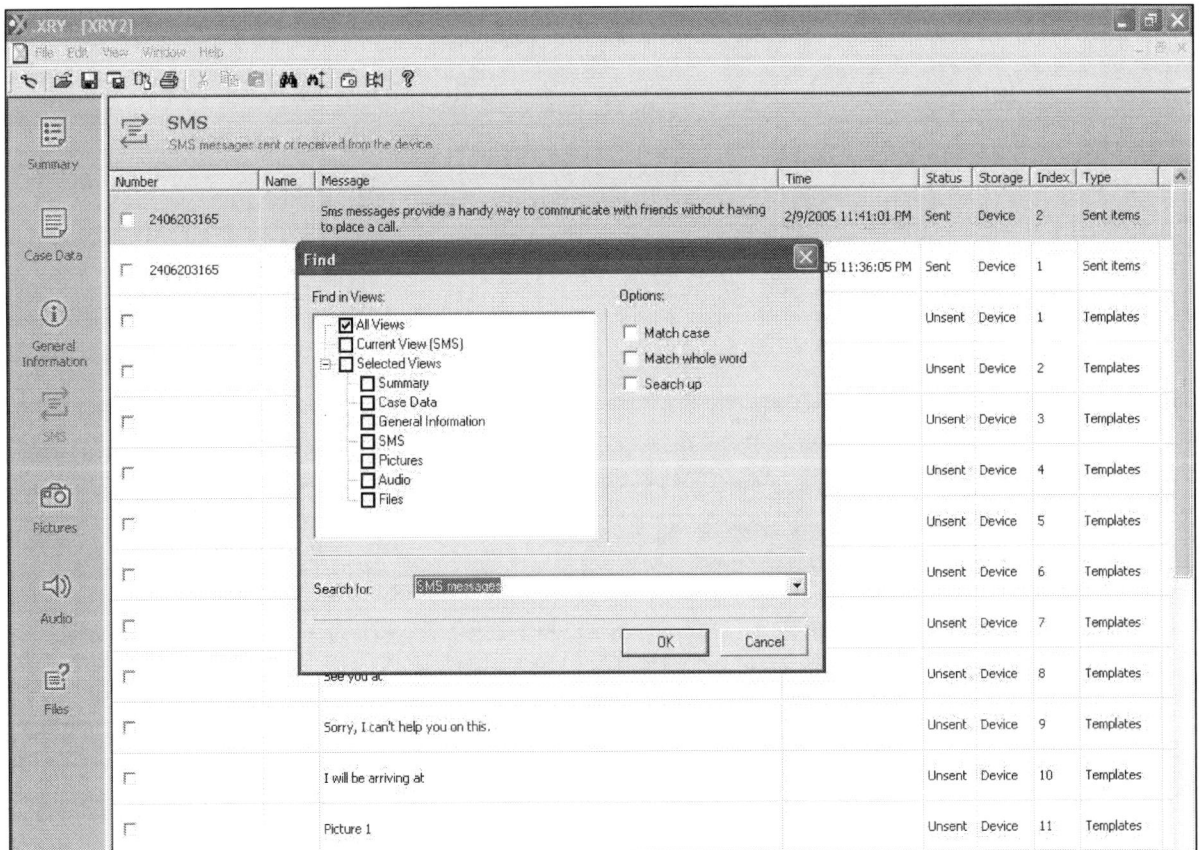

Figure 32: File Content String Search

Graphics Library

The graphics library enables examiners to examine the collection of graphics files present on the device. Each image present can be viewed internally with the Picture Window application, allowing examiners to enlarge images if necessary. Additionally, images collected can be exported and inspected with a third party tool, if necessary. Figure 33 shows a screen shot of images acquired from a Nokia 6610i.

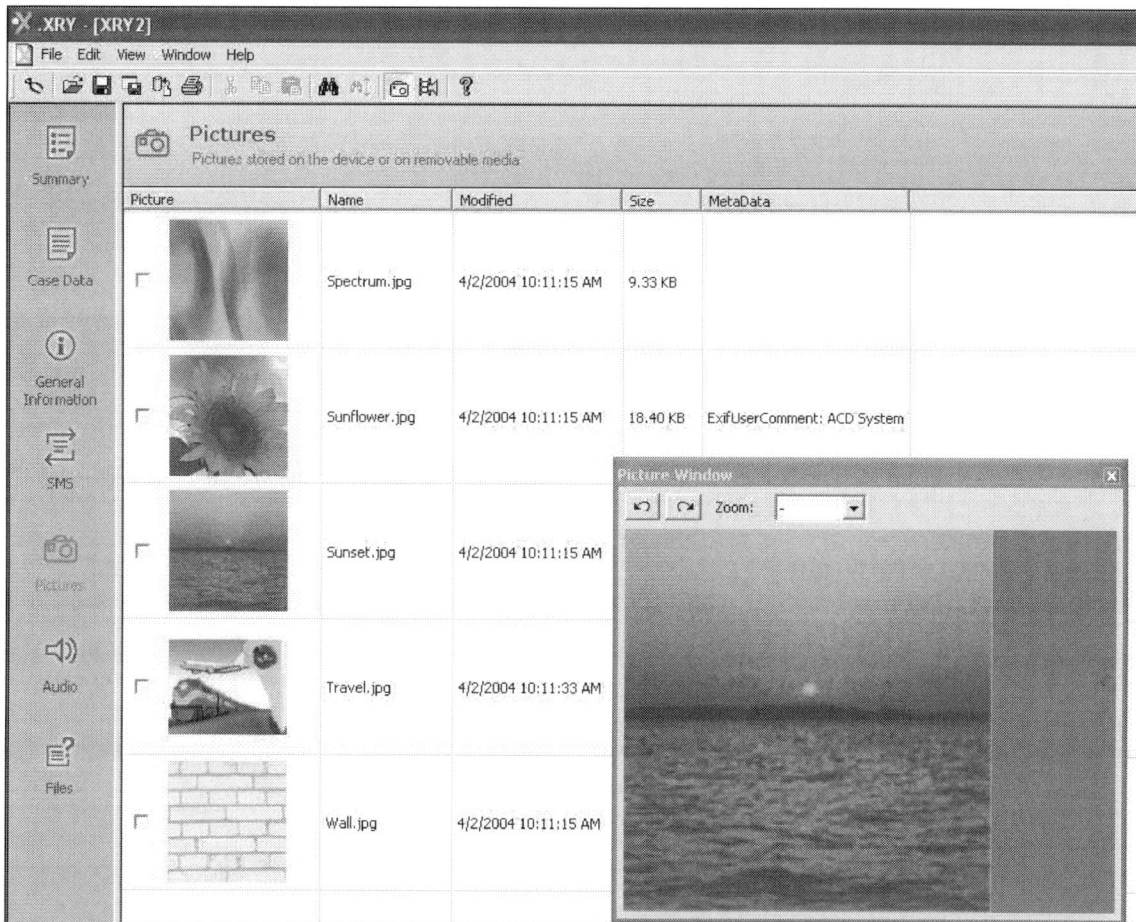

Figure 33: Graphics Library

Report Generation

GSM .XRY allows customized reports to be created with predefined data selection, as illustrated in Figure 34. Bookmarking facilities do not exist in GSM .XRY. Therefore examiners cannot filter data within selected categories. Additionally, reports do not include an illustrative view of acquired graphics files; only filenames, file size, and meta-data are included.

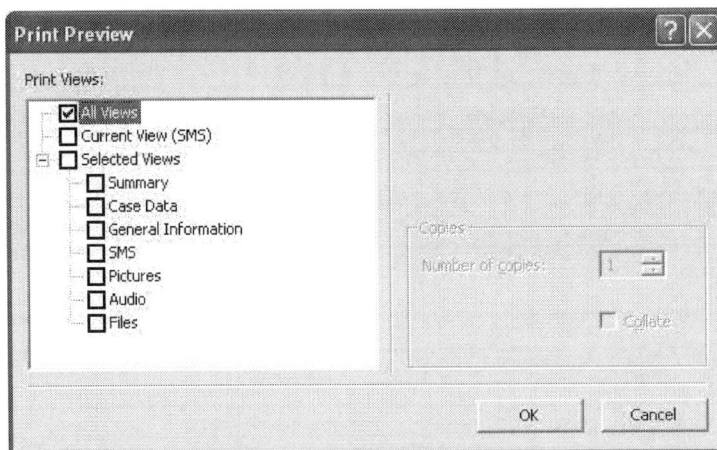

Figure 34: Report Generation

As stated above, examiners have the ability to include all data acquired from the cell phone or to choose a particular category of information. Illustrated below in Figure 35 and Figure 36 are snapshots of two reports generated when choosing respectively the General Information view and a Pictures view.

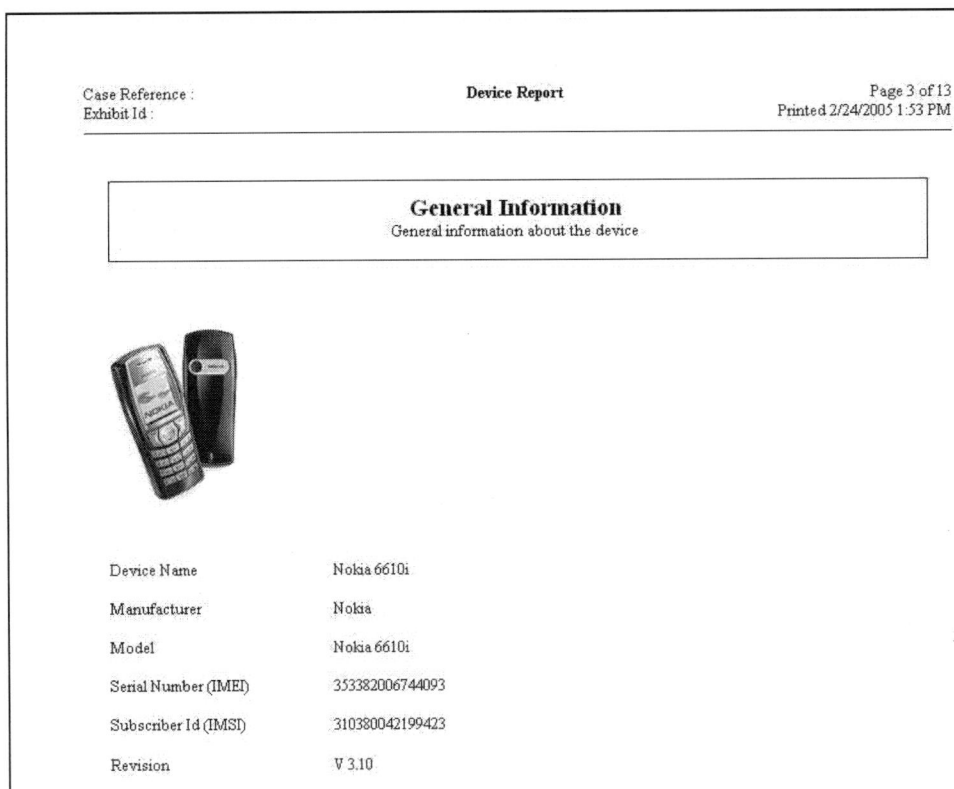

General Information
General information about the device

Device Name	Nokia 6610i
Manufacturer	Nokia
Model	Nokia 6610i
Serial Number (IMEI)	353382006744093
Subscriber Id (IMSI)	310380042199423
Revision	V 3.10

Figure 35: Report Excerpt (General Information)

Pictures
Pictures stored on the device or on removable media

Name: Beach.jpg
Modified: 4/2/2004 10:11:14 AM
Size: 11.42 KB
MetaData: ExifUserComment: ACD Systems Digital Imaging

Name: Blueball.jpg
Modified: 4/2/2004 10:11:15 AM
Size: 7.46 KB
MetaData: ExifUserComment: ACD Systems Digital Imaging

Name: Clip-art01.gif
Modified: 4/2/2004 10:11:33 AM
Size: 1019 Bytes
MetaData: FrameDelay: 0

Name: Clip-art02.gif
Modified: 4/2/2004 10:11:33 AM
Size: 912 Bytes
MetaData: FrameDelay: 0

Name: Clip-art03.gif
Modified: 4/2/2004 10:11:33 AM
Size: 966 Bytes
MetaData: FrameDelay: 0

Name: Clip-art04.gif
Modified: 4/2/2004 10:11:33 AM
Size: 1.04 KB
MetaData: FrameDelay: 0

Name: Clip-art05.gif
Modified: 4/2/2004 10:11:33 AM
Size: 1.02 KB
MetaData: FrameDelay: 0

Figure 36: Report Excerpt (Pictures)

Scenario Results

Table 13 summarizes the results from applying the scenarios listed at the left of the table to the devices across the top. More information can be found in Appendix D: GSM .XRY Results.

Table 13: Results Matrix

Scenario	Device					
	Ericsson T68i	Motorola V66	Motorola V300	Nokia 6610i	Nokia 6200	Nokia 7610
Connectivity and Retrieval	Meet	Meet	Meet	Meet	Meet	Meet
PIM Applications	Below	Below	Below	Below	Below	Below
Dialed/Received Phone Calls	Below	Below	Below	Below	Below	Miss
SMS/MMS Messaging	Below	Below	Below	Below	Below	Miss
Internet Messaging	Miss	NA	Below	NA	NA	Miss
Web Applications	Miss	NA	Miss	Below	Miss	Miss
Text File Formats	NA	NA	Miss	Below	Below	Below

Scenario	Device					
	Ericsson T68i	Motorola V66	Motorola V300	Nokia 6610i	Nokia 6200	Nokia 7610
Graphics File Formats	Miss	NA	Miss	Below	Below	Below
Compressed Archive File Formats	NA	NA	NA	Meet	Meet	Miss
Misnamed Files	NA	NA	Miss	Meet	Meet	Meet
Peripheral Memory Cards	NA	NA	NA	NA	NA	Below
Acquisition Consistency	NA	NA	NA	NA	NA	NA
Cleared Devices	Meet	Meet	Above	NA	NA	Meet
Power Loss	Above	Above	Above	Above	Above	Above

SIM Card Acquisition

GSM .XRY version 2.4 through 2.5 allows examiners the ability to acquire data directly from a SIM card using the CardMan reader by OMNIKEY. The acquisition steps followed to acquire data directly from the SIM are the same as acquiring data from a phone, except for selecting SIM/USIM Card from the user interface in lieu of Mobile Phone. The data fields acquired (e.g., General Information (i.e., ICCID, IMSI, Phase), Contacts, Calls, Messages, etc.) are dependent upon the SIM and service provider. The Search engine and Report facilities operate in a similar fashion as with phone acquisitions, described above.

Table 14 summarizes the results from applying the scenarios listed at the left of the table to the SIMs across the top. More information can be found in Appendix O: GSM .XRY – External SIM Results.

Table 14: SIM Card Results Matrix - External Reader

Scenario	SIM		
	5343	8778	1144
Basic Data	Below	Below	Below
Location Data	Below	Below	Below
EMS Data	Meet	Meet	Meet
Foreign Language Data	Below	Below	Below

Synopsis of Oxygen Phone Manager

Oxygen Phone Manager (OPM)[13] is a tool designed to manage information on a cell phone, including contacts, calendar, SMS messages, to-do list, logs, and ring tones. The software is designed to support most Nokia phones. A Symbian OS version also is available. Oxygen Phone Manager - Forensic Version is an adaptation of the phone management tool that suppresses changing data on the phone, but allows data to be logically acquired and exported into several supported formats. Version 2.6 (Build 0.1) of the program works under Microsoft Windows 98, ME, 2000, and XP operating systems.

Supported Phones

The OPM forensic version, like the regular version is targeted for Nokia phones.[14] However, the tool does not support Nokia Symbian OS smart phones. The tool is designed to acquire phonebook contacts (including pictures), Call Lists (i.e., last numbers dialed, missed and received calls), SMS messages, pictures, logos, ring tones, profiles, To-Do lists, MMS messages (supported formats are plain text, HTML, JPEG, GIF, animated GIF, PNG, TIFF, BMP, MIDI, WAV, and RT), java applications, games, gallery and play lists.

Acquisition Stage

The phone memory is acquired along with some SIM-resident information, such as phonebook entries. The tool is not designed for SIM card extraction via an external smart card or SIM reader. The acquisition process is conducted through a wizard, as illustrated below in Figure 37. First, either connection parameters can be pre-configured via the program options or the type of connection to use must be selected among cable, infrared, and Bluetooth options.

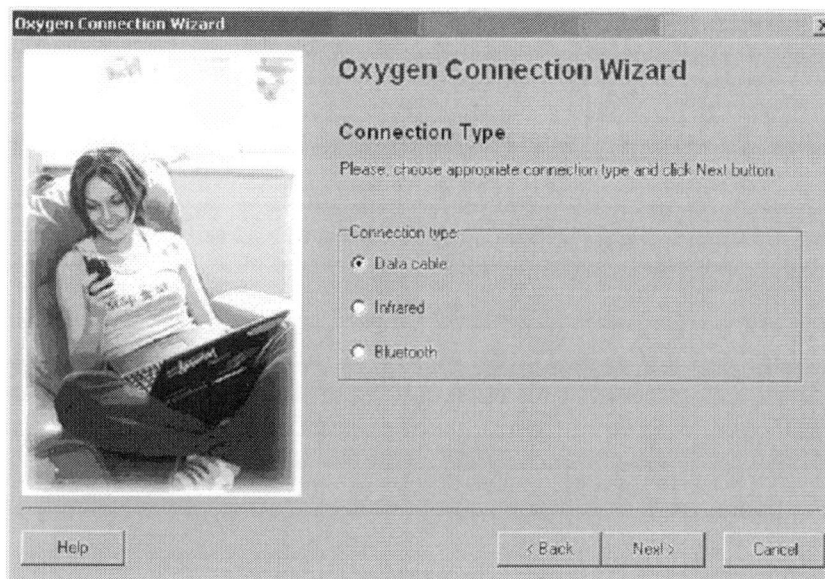

Figure 37: Interface Selection

---

[13] Additional information can be found at: **http://www.opm-2.com/forensic**
[14] Additional information on supported phone models can be found at: **http://www.oxygensoftware.com/products/opm2**

Then, the correct configuration (i.e., type of cable) for the connection may be selected, if necessary.

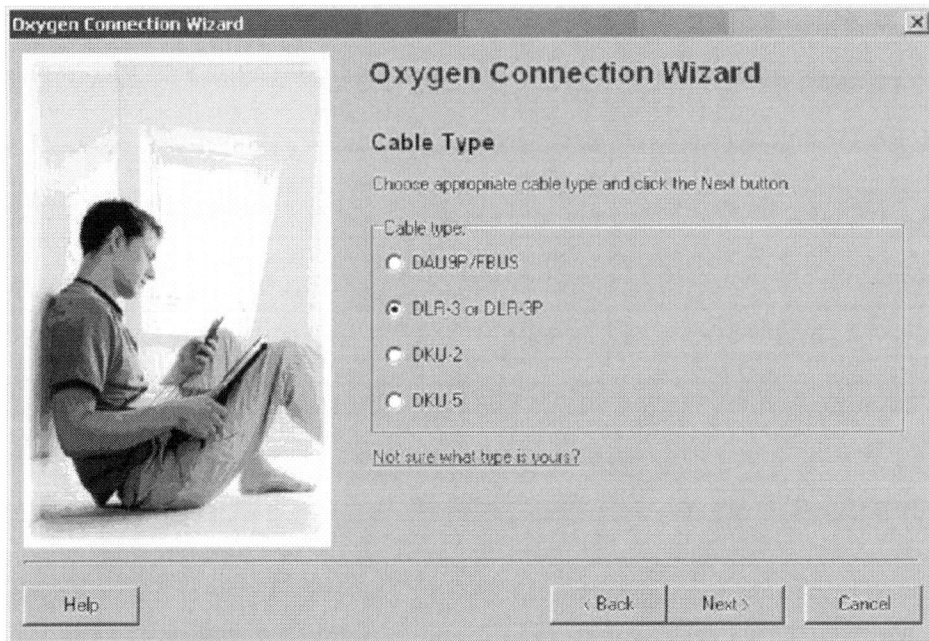

Figure 38: Cable Interface Selection

After the proper interface selection is made, the connectivity between the forensic workstation and the device is queried, as illustrated below in Figure 39.

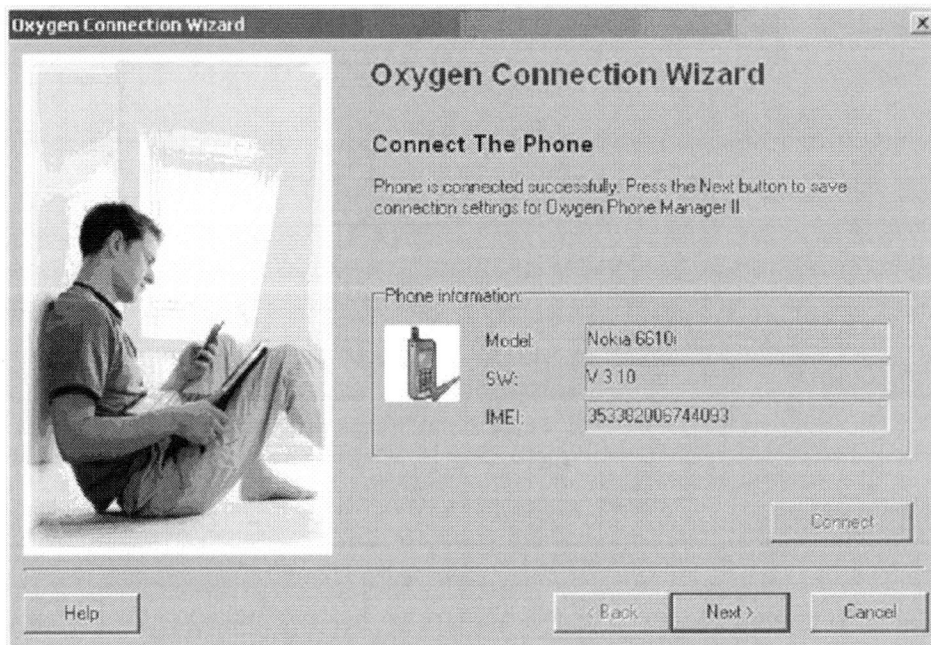

Figure 39: Device Identification

Once the configuration wizard is finished, the main application can be launched to detect the phone and download all the selected information from it.

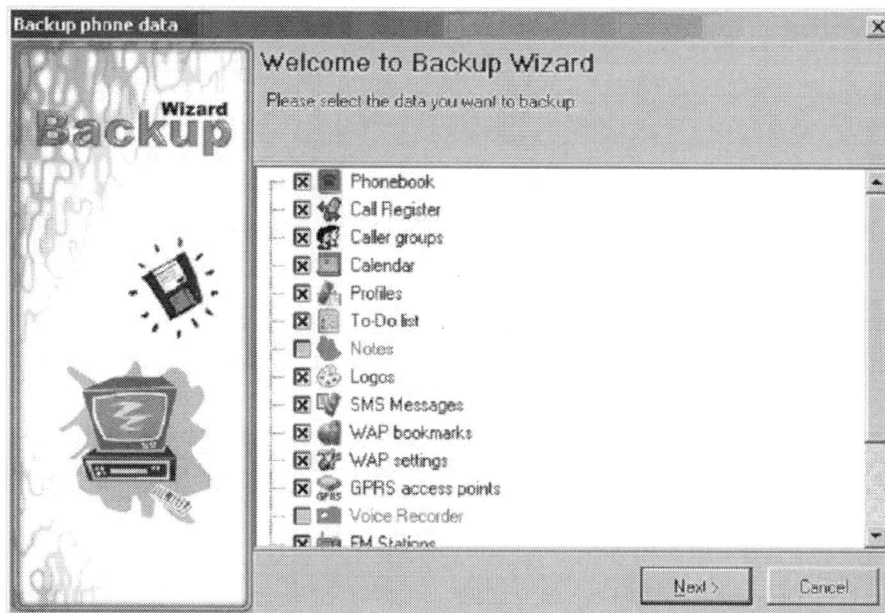

Figure 40: Data Acquisition Selection

Search Functionality

The forensic version of OPM provides no search functionality. However, a tree structure of the acquisition is built and populated with the selected data. All the items can be saved or exported in common formats and searched using a third party tool.

Graphics Library

The logos and graphic files are retrieved by the tool via the tree structure at the left of the phone content explorer screen, as illustrated in Figure 41.

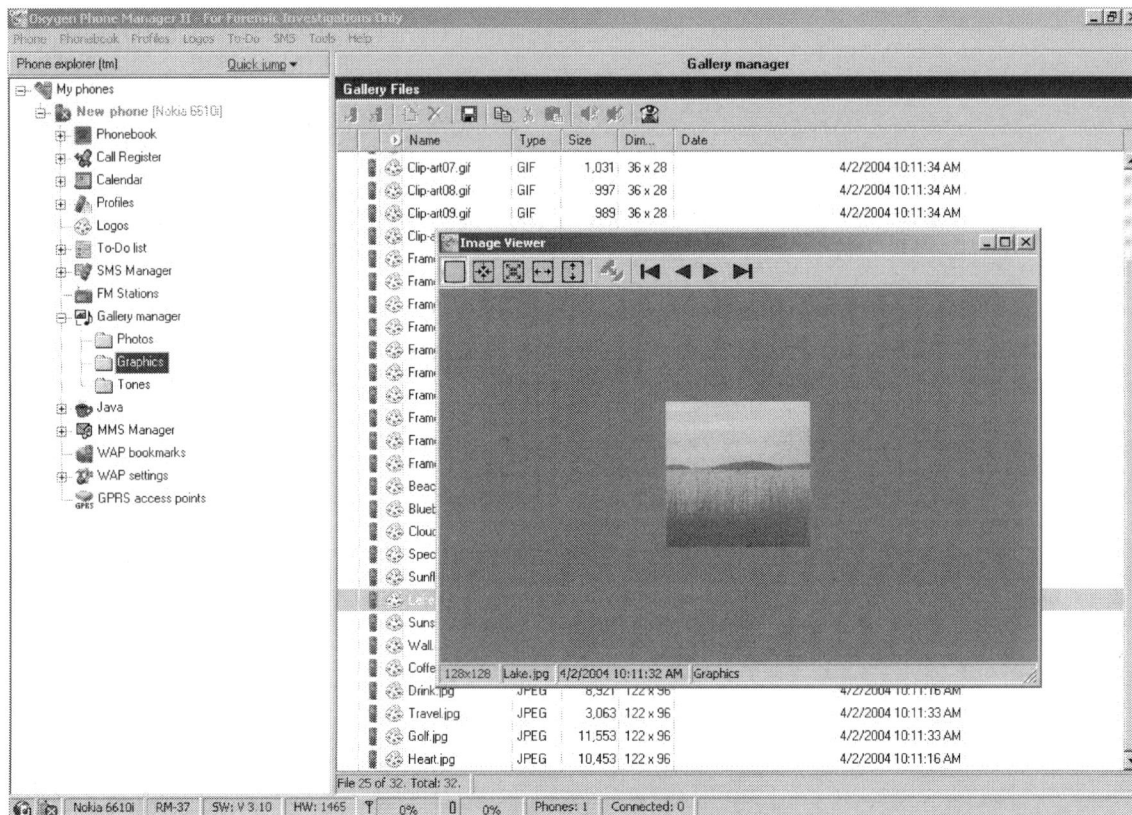

Figure 41: Graphics Library

Report Generation

No report generation facility is provided. The data extracted from the phone through the tool interface must be exported manually and processed through another means, such as a word processor. Data can be saved in several different formats (i.e., .rtf, .html, .csv, etc.) for export. Certain data, such as Profiles, can only be saved in the OPM proprietary data format.

Scenario Results

Table 15 summarizes the results from applying the scenarios listed at the left of the table to the devices across the top. More information can be found in Appendix E: Oxygen Phone Manager Results.

Table 15: Results Matrix

Scenario	Device			
	Nokia 3390	Nokia 6610i	Nokia 6200	Nokia 7610
Connectivity and Retrieval	Meet	Meet	Meet	Meet
PIM Applications	Below	Below	Below	Below
Dialed/Received Phone Calls	Below	Below	Below	Miss
SMS/MMS Messaging	Below	Below	Below	Below
Internet Messaging	NA	NA	NA	Below

Scenario	Device			
	Nokia 3390	Nokia 6610i	Nokia 6200	Nokia 7610
Web Applications	Miss	Miss	Miss	Miss
Text File Formats	NA	Below	Below	Miss
Graphics File Formats	NA	Below	Below	Below
Compressed Archive File Formats	NA	Meet	Meet	Miss
Misnamed Files	NA	Meet	Meet	Miss
Peripheral Memory Cards	NA	NA	NA	Below
Acquisition Consistency	NA	NA	NA	NA
Cleared Devices	NA	NA	NA	Meet
Power Loss	Above	Above	Above	Above

Synopsis of MOBILedit!

MOBILedit! Forensic, version 1.93, is able to acquire information from various GSM, CDMA, PCS cell phones from assorted manufacturers (i.e., Alcatel, Ericsson, General, LG, Motorola, Nokia, Panasonic, Philips, Samsung, Siemens, Sony Ericsson).[15] A non-forensic variant of the product, called MOBILedit!, also exists. This report covers only the forensic version, which for simplicity is referred to as MOBILedit!. The MOBILedit! application provides the ability to perform acquisitions via cable, IrDA (Infrared), or Bluetooth interfaces.

Supported Phones

The data acquired using MOBILedit! depends upon the make, model, and richness of phone features. Some common data fields acquired using MOBILedit! are phone and subscriber information, Phonebook, SIM Phonebook, Missed calls, Last Numbers Dialed, Received calls, Inbox, Sent items, Drafts, and Files (e.g., Graphics, Photos, Tones). Figure 42 shows a screen shot of the internally listed phones. The settings selected allow the examiner to identify the specific manufacturer of the device before acquisition.

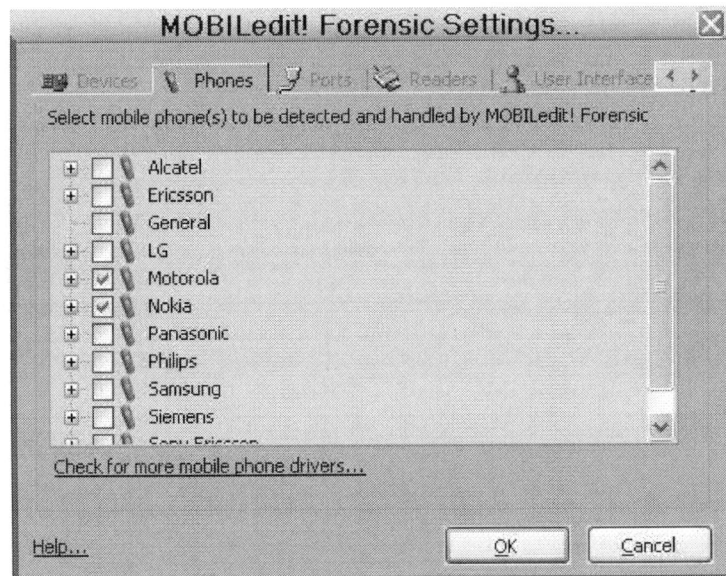

Figure 42: Supported Phone Selection

Acquisition Stage

As mentioned earlier, cell phone devices can be acquired using either an appropriate cable, IrDA, or Bluetooth interface. To create an interface between the forensic workstation and the device, examiners must select the type of connection to use, as illustrated in Figure 43. For a cable connection, the examiner must identify the port being used by the phone, as illustrated below in Figure 44. Upon a successful connection MOBILedit! recognizes the device and automatically begins logical acquisition. If the device is not recognized, the examiner has the option of identifying the connected device manually through the MOBILedit! wizard.

---

[15] Additional information on supported phone models can be found at: **www.mobiledit.com**

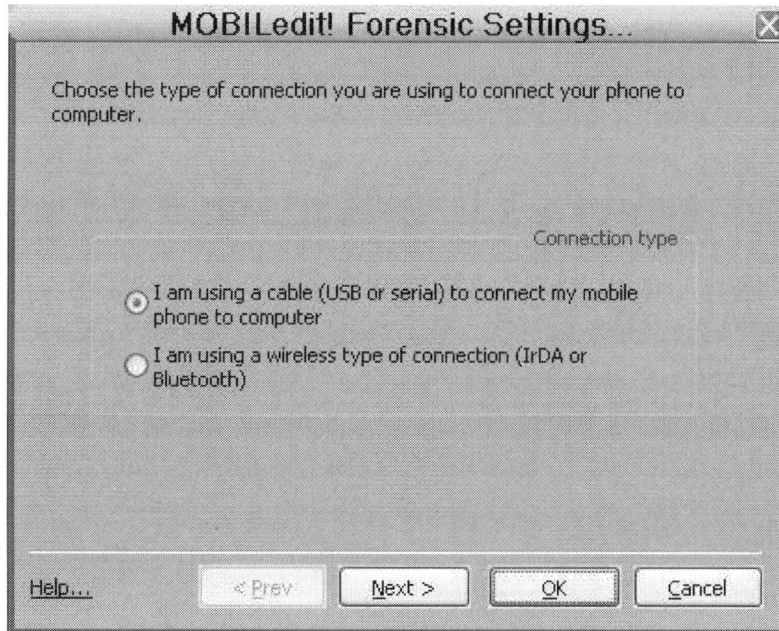

Figure 43: Interface Selection

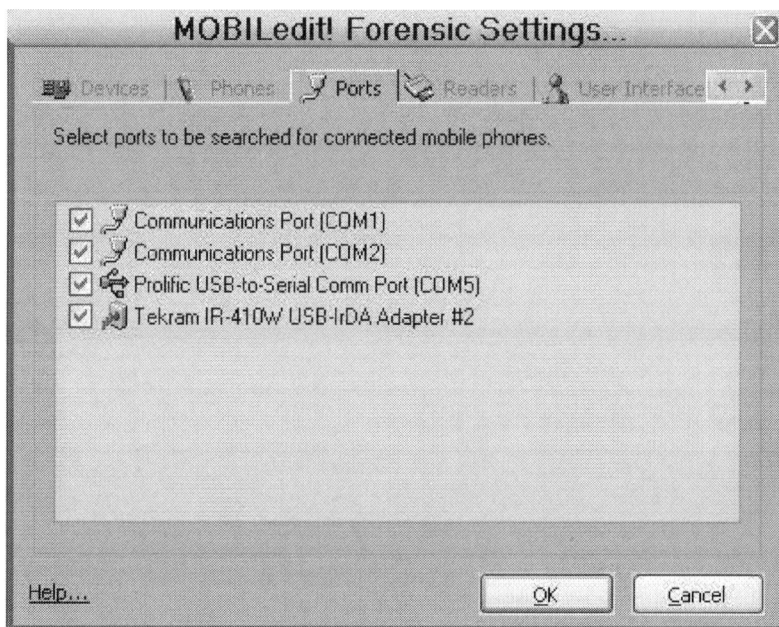

Figure 44: Port Selection

The MOBILedit! phone content interface appears after the acquisition completes, as illustrated below in Figure 45. The window is divided into two major panes, allowing the examiner to navigate and examine the logically acquired data.

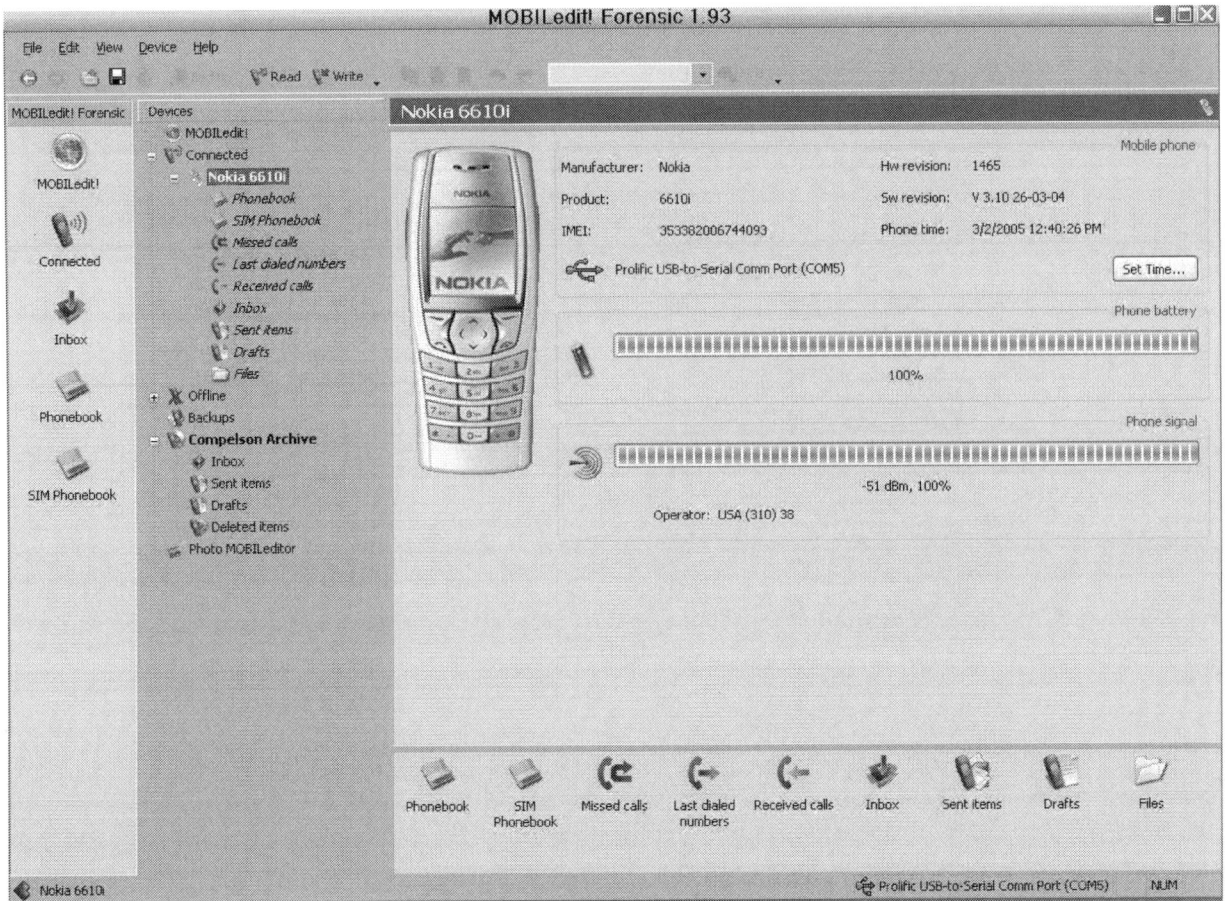

Figure 45: MOBILedit! User Interface

MOBILedit's search engine allows examiners to perform simplified string searches only within specific folders. The search engine does not give examiners the ability to search through multiple cases, multiple folders within a case, or issue complex expression patterns.

Graphics Library

The graphics library enables examiners to examine the collection of graphics files present on the device. Each image present can be viewed internally with the Picture Window application allowing examiners to enlarge images if necessary. Additionally, images collected can be exported and inspected with a third party tool, if necessary. Figure 46 shows a screen shot of images acquired, while Figure 47 illustrates the single picture view that is provided when a single image is selected.

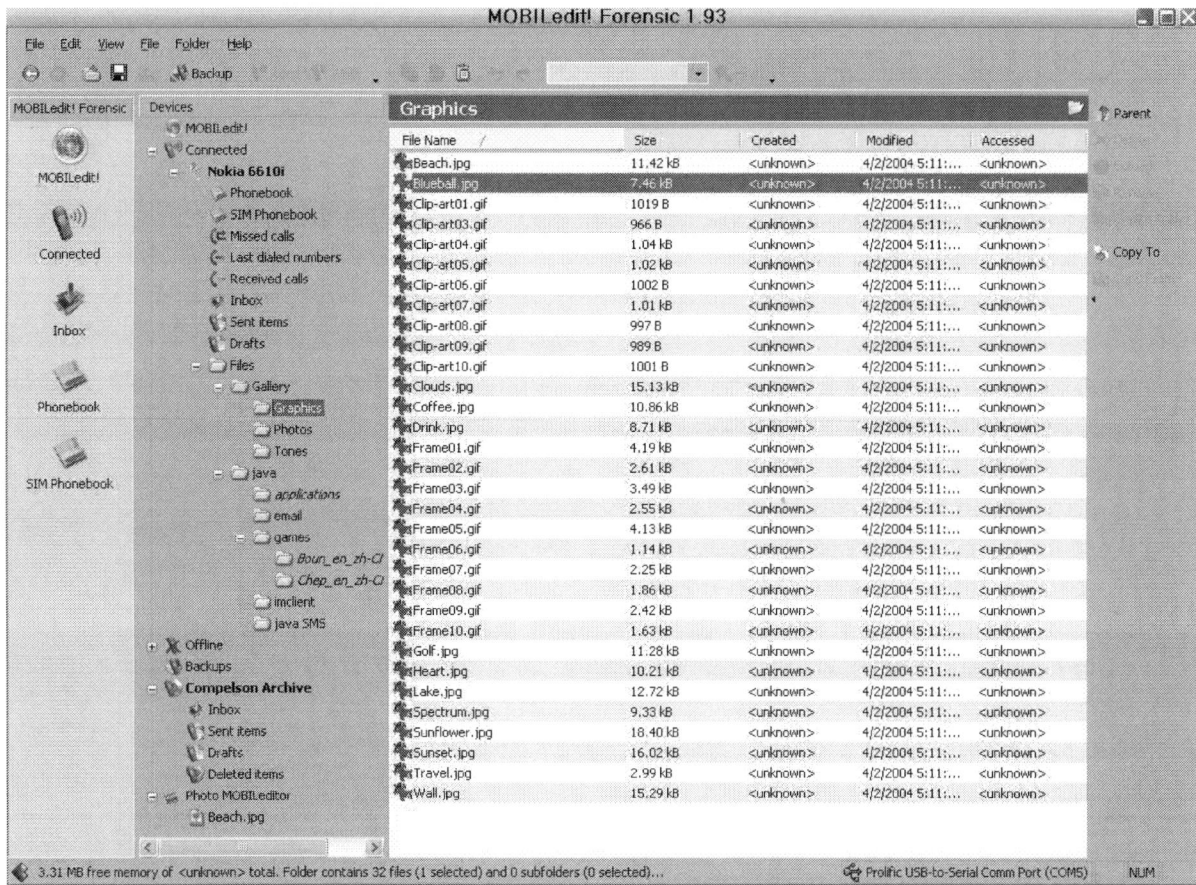

Figure 46: Graphics Library (File Listing)

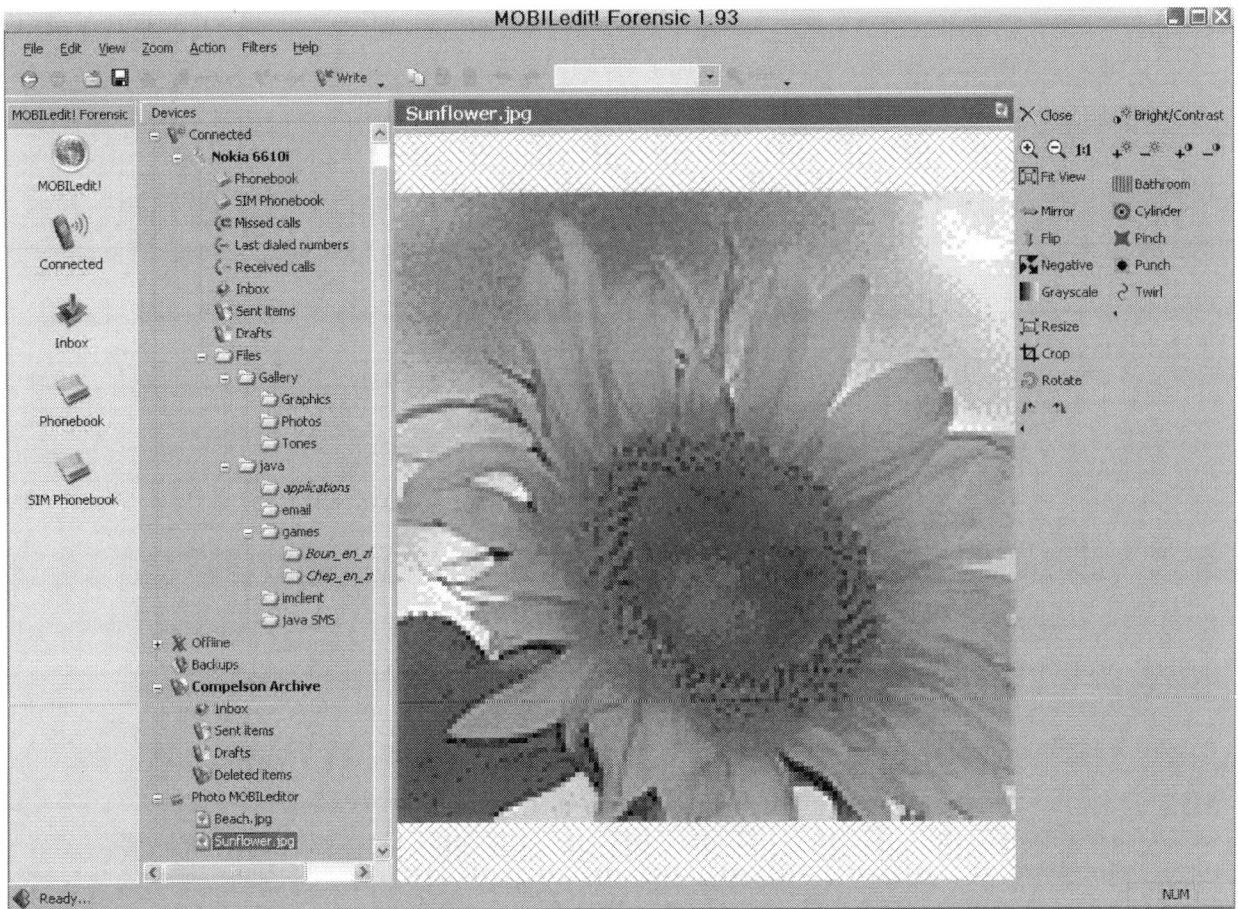

Figure 47: Graphics Library (Single Picture)

Report Generation

MOBILedit! versions before 1.95 do not have a reporting facility. Relevant data collected from the device can be copied over to another, by right clicking on a specified item and pasting into the destination document. A document editor can then be used to create a finalized report of significant findings. More recently, however, version 1.95 and above incorporate report generation, allowing examiners to produce reports internally within the application or to export them in an xml format. Figure 48 below illustrates the interface for report generation.

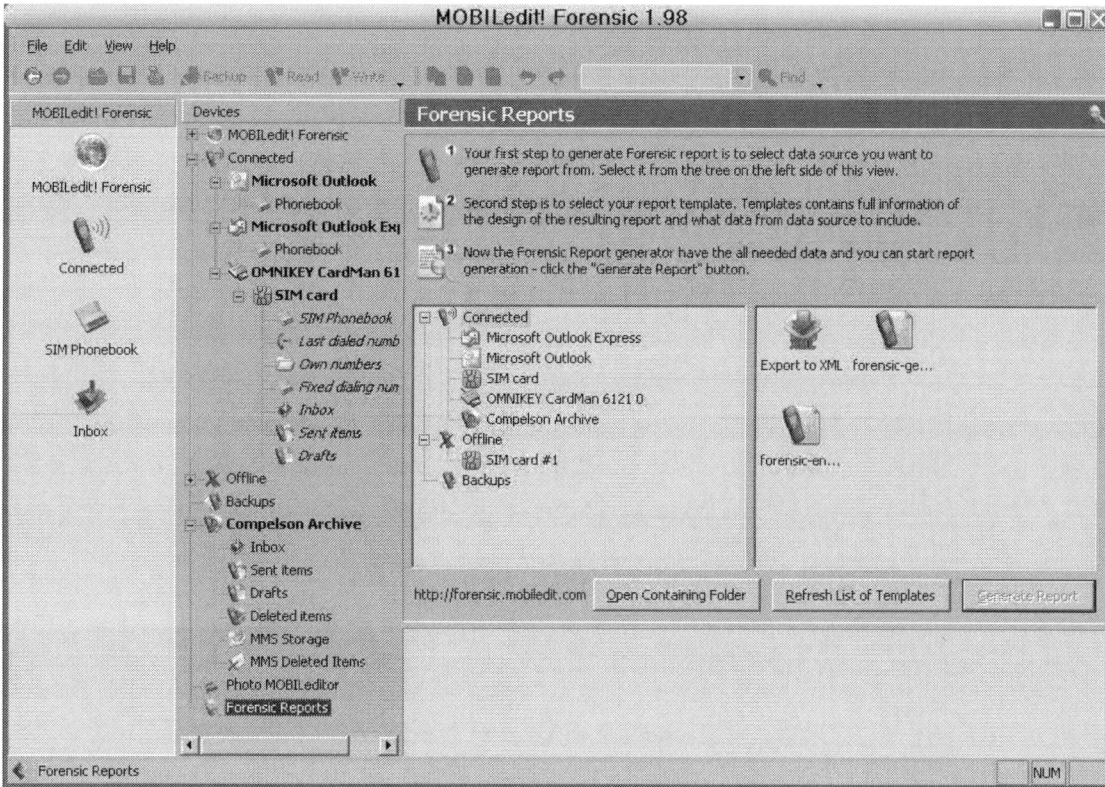

Figure 48: Report Generation

A screen shot representing the internal report output is illustrated below in Figure 49.

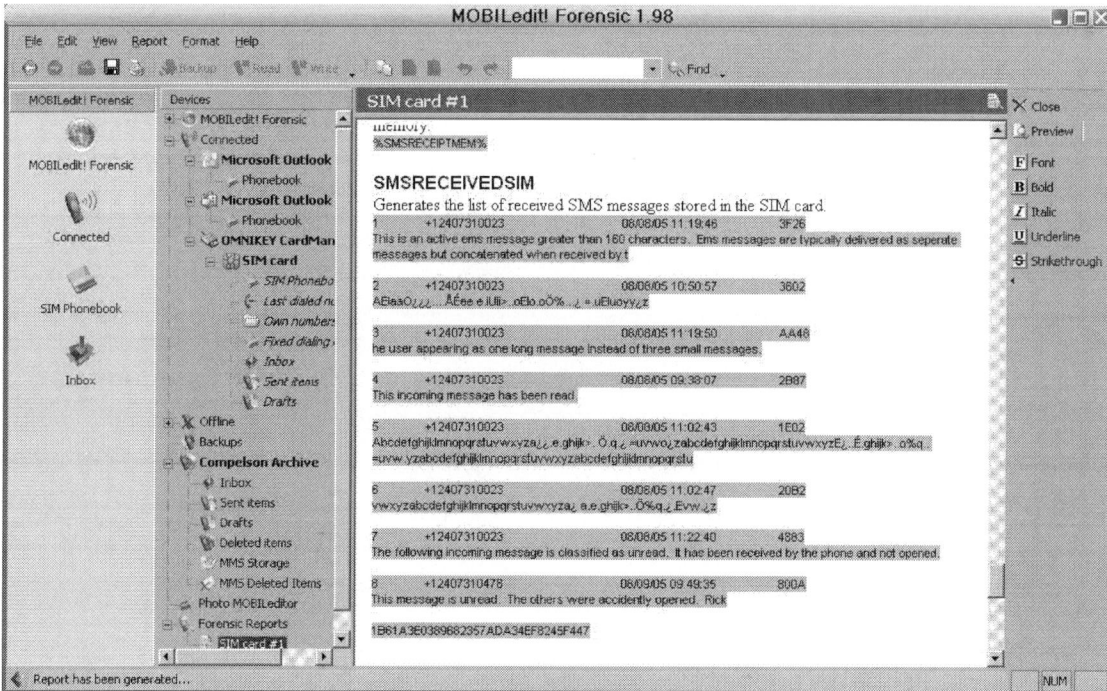

Figure 49: Internal Report Format

Figure 50 shows the user interface, which allows the examiner to select specific fields to be included in the XML-generated report displayed in Figure 51.

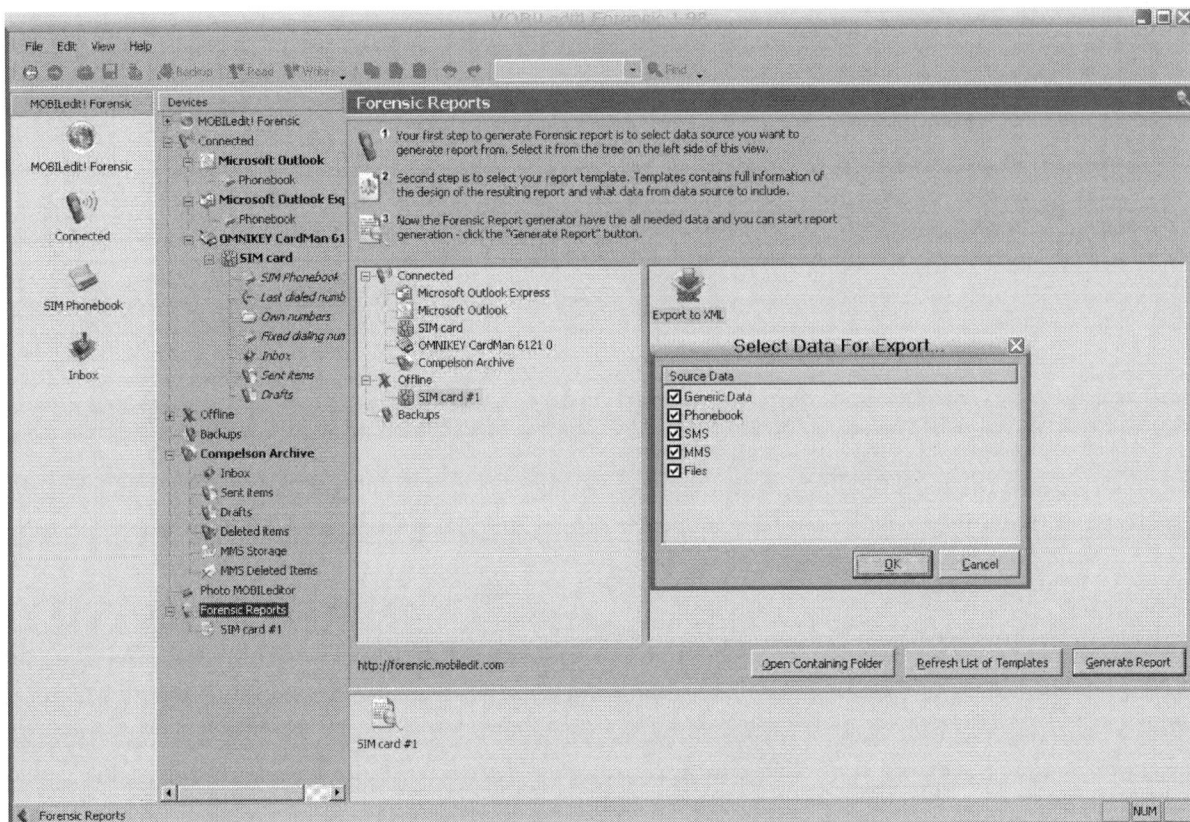

Figure 50: XML Report Generation

Figure 51: XML Report Excerpt

Table 16 summarizes the results from applying the scenarios listed at the left of the table to the devices across the top. More information can be found in Appendix F: MOBILedit! Results.

Table 16: Results Matrix

Scenario	Device				
	Ericsson T68i	Motorola C333	Motorola V66	Motorola V300	Nokia 6610i
Connectivity and Retrieval	Meet	Meet	Meet	Meet	Meet
PIM Applications	Miss	Below	Below	Below	Below
Dialed/Received Phone Calls	Below	Below	Below	Below	Below
SMS/MMS Messaging	Below	Miss	Miss	Below	Below
Internet Messaging	Miss	NA	NA	Below	NA
Web Applications	Miss	NA	NA	Miss	Miss
Text File Formats	NA	NA	NA	Miss	Below
Graphics File Formats	Miss	NA	NA	Miss	Below
Compressed Archive File Formats	NA	NA	NA	NA	Meet
Misnamed Files	NA	NA	NA	NA	Meet
Peripheral Memory Cards	NA	NA	NA	NA	NA
Acquisition Consistency	NA	NA	NA	NA	NA
Cleared Devices	Meet	Above	Meet	Above	NA
Power Loss	Above	Above	Above	Above	Above

Mobiledit! Forensic version 1.93 and above gives examiners the ability to acquire SIM card data using a PC/SC-compatible reader. The acquisition steps followed to acquire data directly from the SIM are the same as acquiring data from a phone, except for selecting Smart Card Readers versus Mobile Phones. The data fields acquired (i.e., SIM Phonebook, Last Numbers Dialed, Fixed Dialing Numbers, Inbox, Sent Items, Drafts) are dependent upon the SIM and service provider. The Search engine and Report facilities operate in a similar fashion as for phone acquisitions, described above.

Table 17 summarizes the results from applying the scenarios listed at the left of the table to the SIMs across the top. More information can be found in Appendix P: Mobiledit! – External SIM Results.

Table 17: SIM Card Results Matrix - External Reader

Scenario	SIM		
	5343	8778	1144
Basic Data	Below	Below	Below
Location Data	Miss	Miss	Miss
EMS Data	Below	Below	Below
Foreign Language Data	Below	Below	Below

Synopsis of BitPIM

BitPIM version 0.7.28 can acquire information from various manufacturers of CDMA cell phones (e.g., Audiovox, Samsung, Sanyo).[16] The BitPIM application provides the ability to perform acquisitions through a cable interface.

Supported Phones

The make, model, and type of CDMA phone determine what data BitPIM can acquire. Some common data fields that BitPIM can recover are: Phonebook, Wallpapers (Graphic files present on the phone), Ringers (sound bytes), Calendar entries, and Memo entries. BitPIM also captures a dump of the filesystem, where data related to Incoming/Outgoing/Missed/Attempted calls and SMS/MMS messages can be found. Figure 52 shows a screen shot of the phones listed in the Settings window under a menu item that allows selection of the specific make and model of the device for acquisition. The Read Only box should always be checked to avoid writing to the phone.

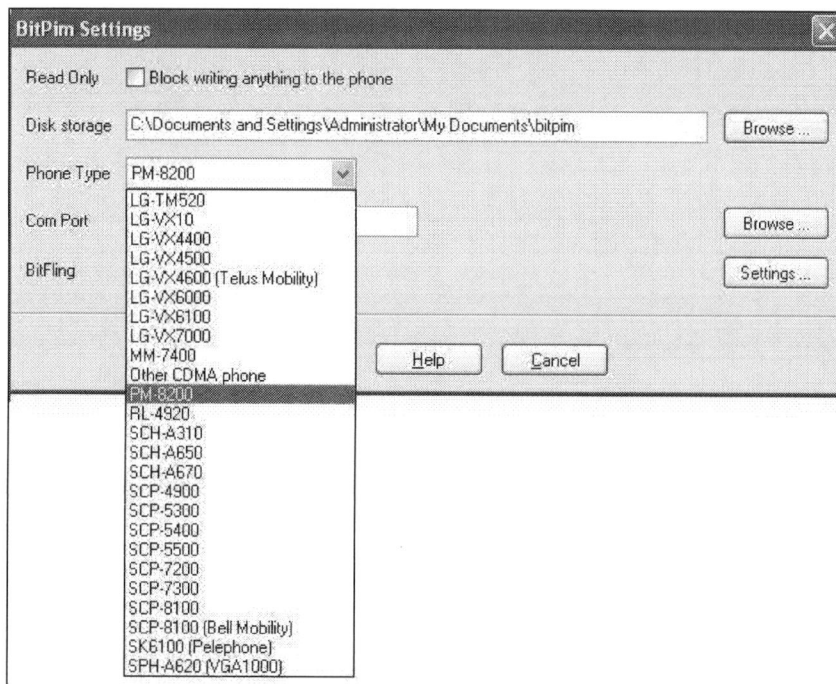

Figure 52: Supported Phone Selection

Acquisition Stage

Because BitPIM acquires data via a cable interface, the proper cable must be used. Susteen's Data Pilot[17] cables were used for phone to PC connectivity. Acquisition begins by selecting the Data -> Get Phone Data menu from the toolbar and specifying which items to acquire, as illustrated below in Figure 53.

---

[16] Additional information on supported phone models can be found at: **www.bitpim.org**
[17] Additional information on DataPilot cables can be found at: **www.datapilot.com**

71

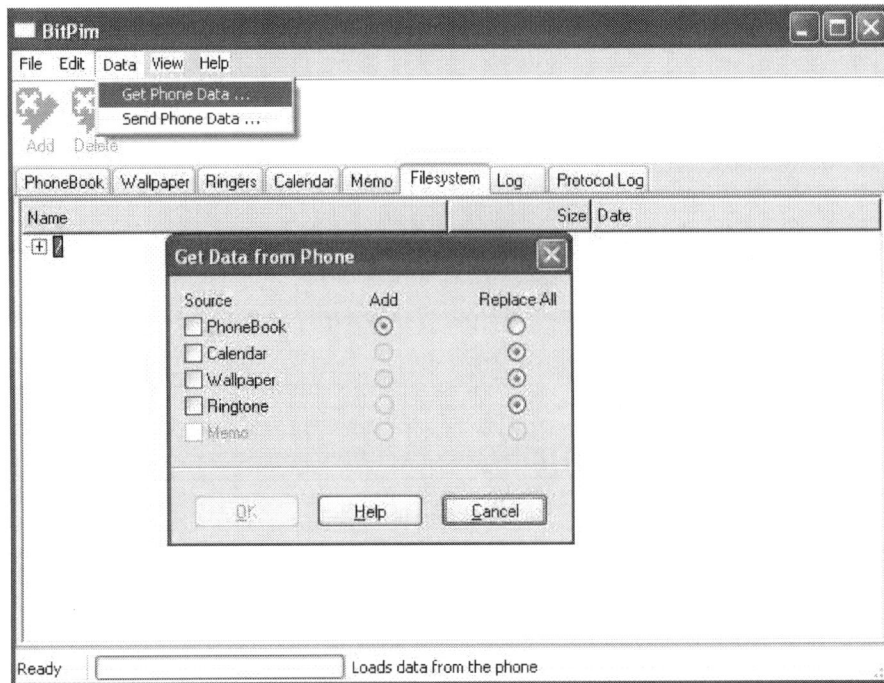

Figure 53: Data Acquisition Selection

Figure 54 and Figure 55 provide sample screen shots after a successful acquisition, displaying active entries from the Contacts and Calendar entries of the cell phone.

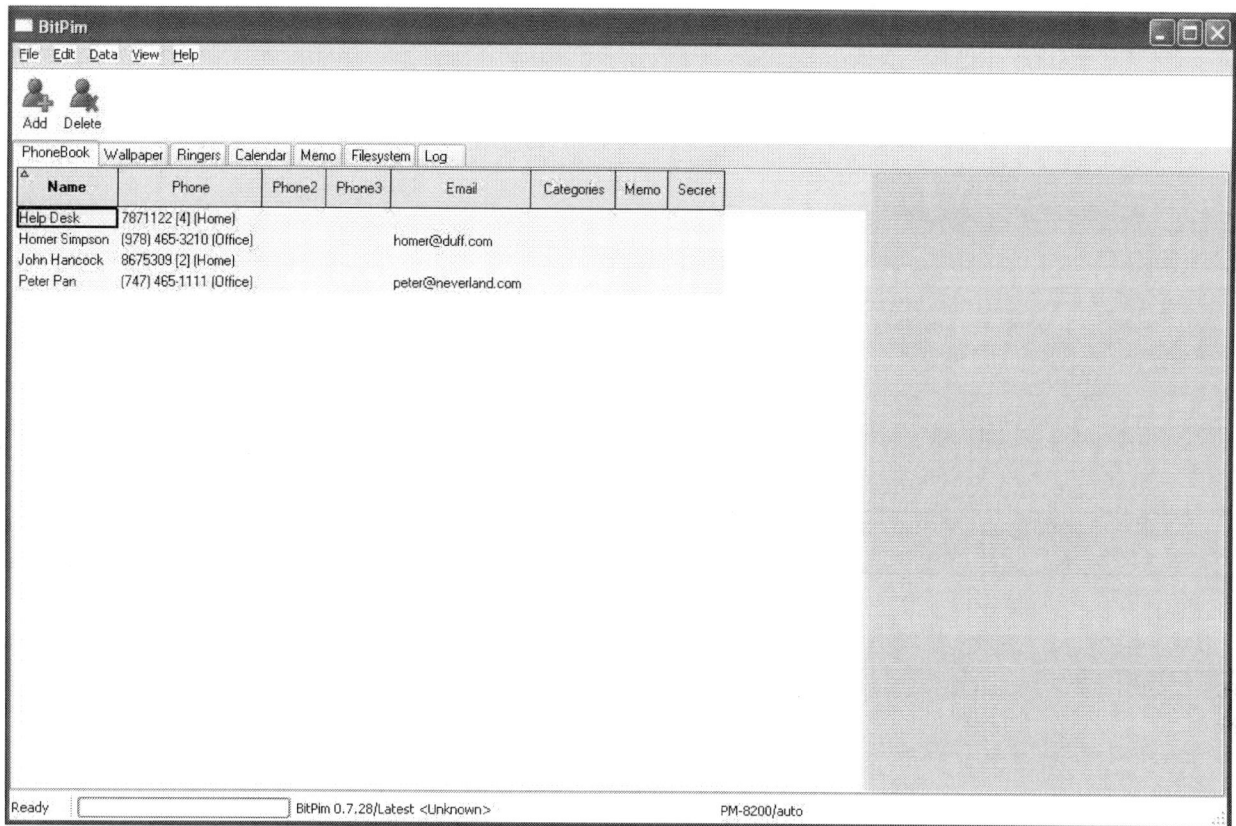

Figure 54: Acquired PhoneBook Data

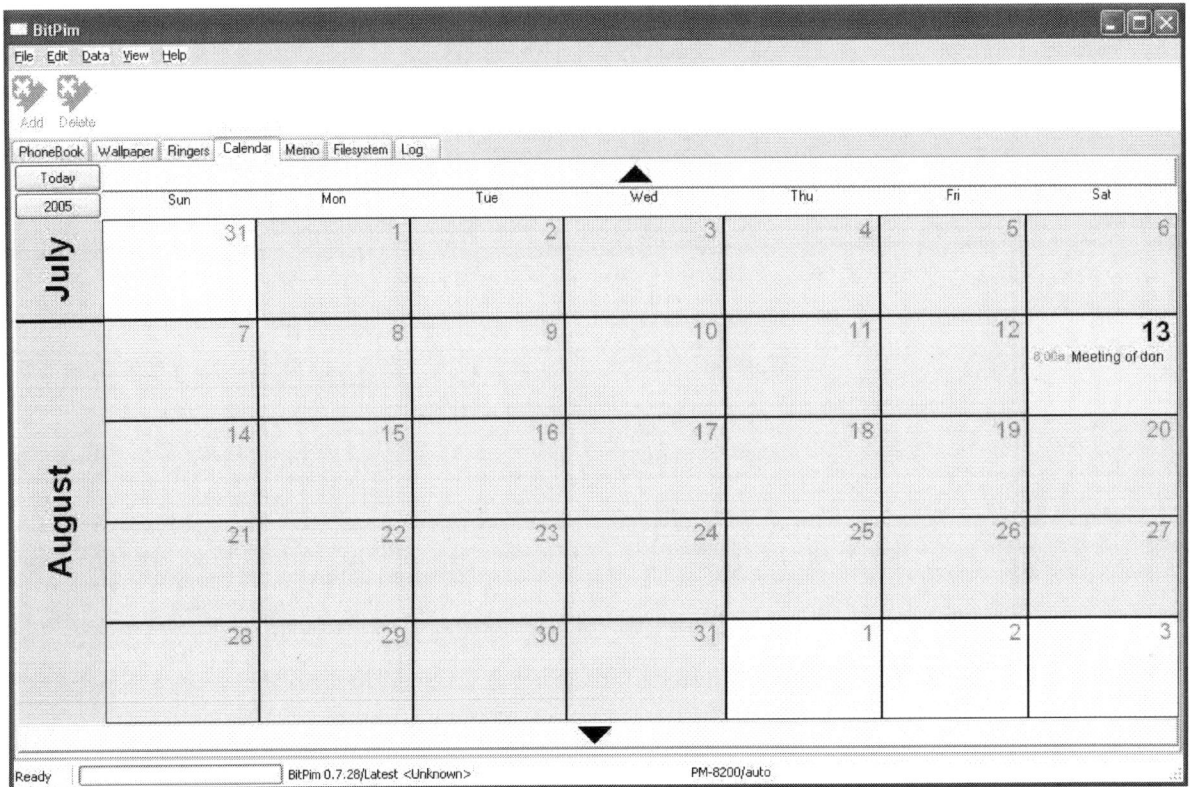

Figure 55: Acquired Calendar Data

BitPIM's filesytem data dump allows examination of SMS, MMS, message content and potential recovery of deleted items related to Phonebook entries and incoming and outgoing messages.

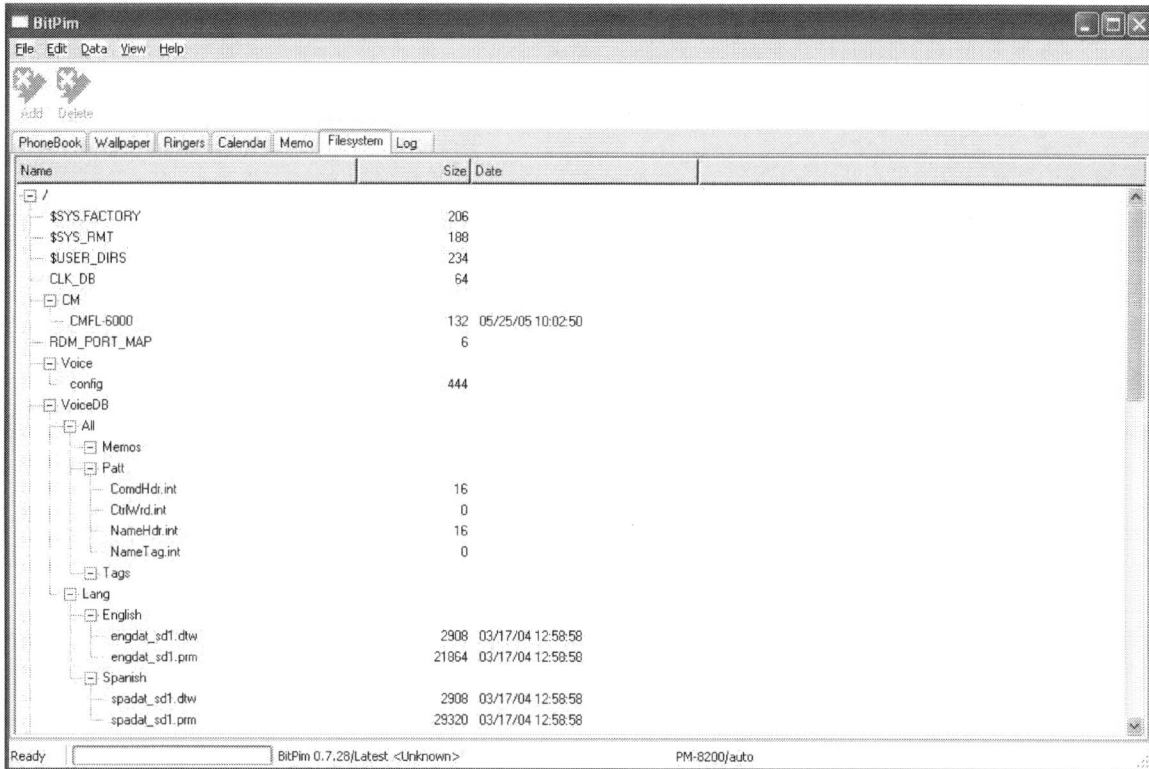

Figure 56: Filesytem Data Dump

Figure 57 provides a screen shot of a recovered deleted phonebook entry.

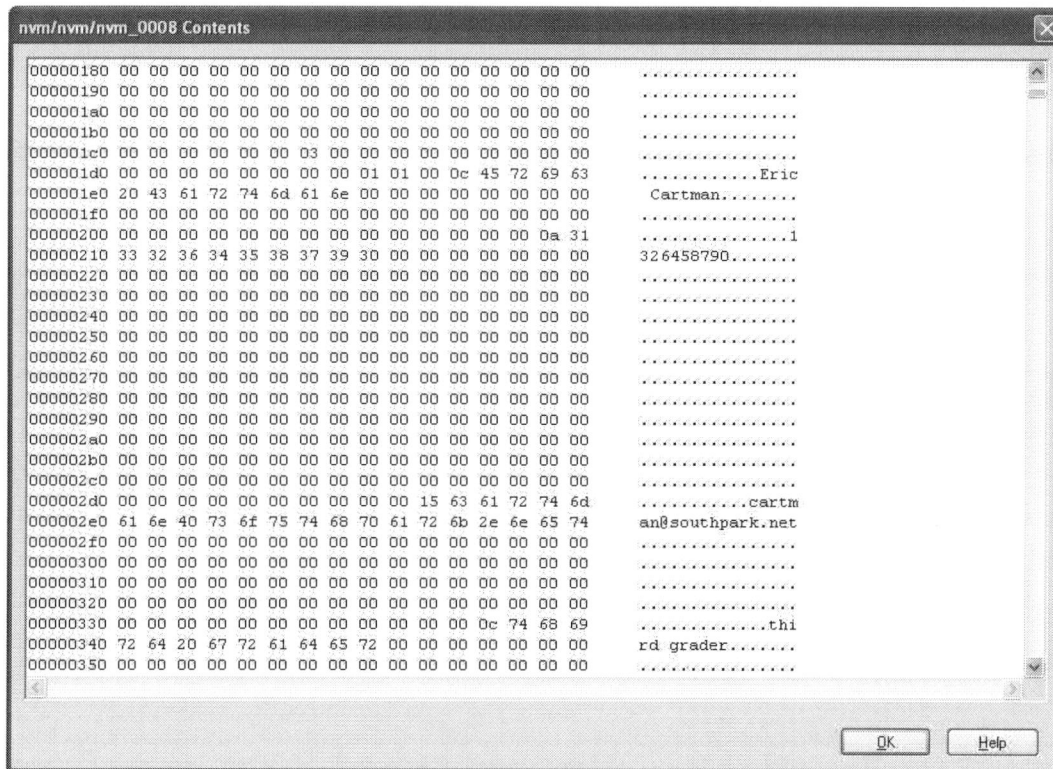

Figure 57: Filesytem Data Dump - Individual File

75

Search Functionality

BitPIM provides no search functionality. However, items can be saved or exported in common formats and searched using a third party tool.

Graphics Library

The Wallpaper tab enables examiners to examine the collection of graphics files present on the device. Each image present can be viewed internally with the Picture Window application allowing examiners to enlarge images if necessary. Additionally, images collected can be exported and inspected with a third party tool, if necessary. Figure 58 shows a screen shot of images acquired.

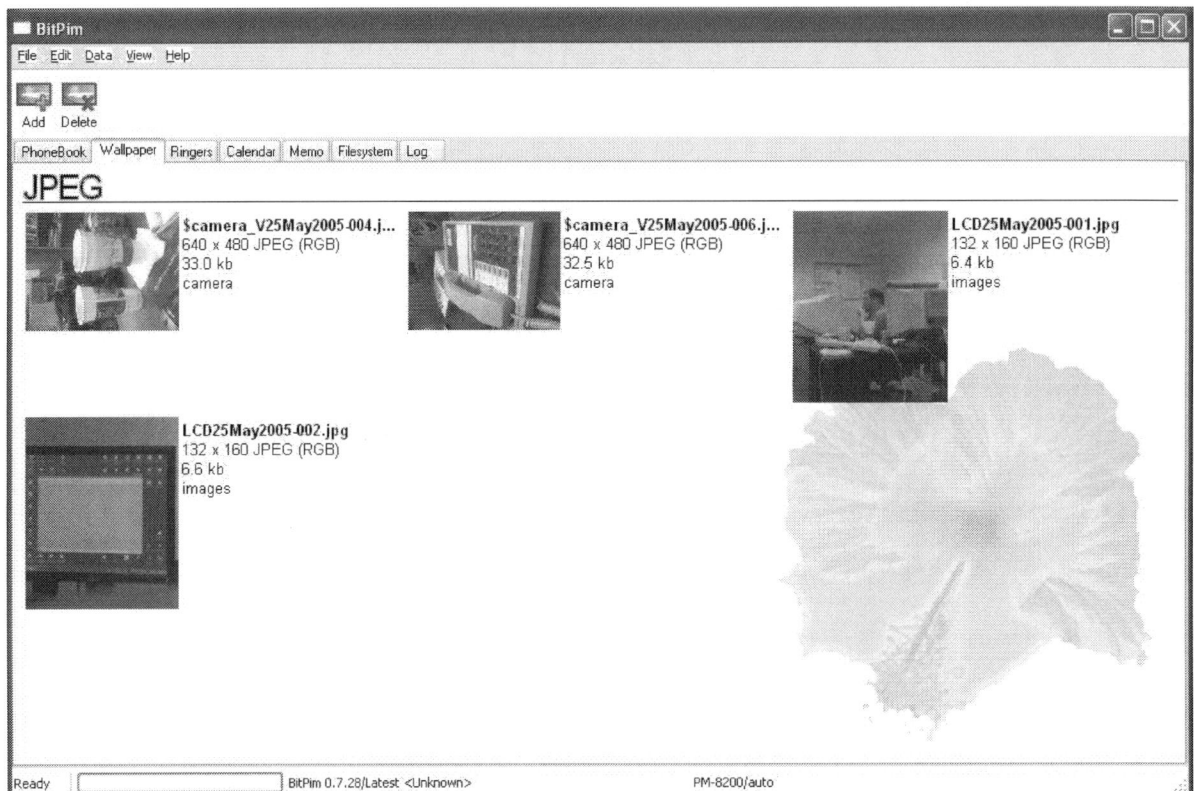

Figure 58: Acquired Wallpaper Data

Report Generation

BitPIM does not support reporting facilities internally. Relevant data collected from the device can be copied, by right clicking on a specified item. Third party tools or editors can be used to create a finalized report of significant findings.

Scenario Results

Table 18 summarizes the results from applying the scenarios listed at the left of the table to the devices across the top. More information can be found in Appendix G: BitPIM Results.

Table 18: Results Matrix

Scenario	Device	
	Audiovox 8910	Sanyo 8200
Connectivity and Retrieval	Meet	Meet
PIM Applications	Meet	Meet
Dialed/Received Phone Calls	Meet	Meet
SMS/MMS Messaging	Below	Meet
Internet Messaging	NA	Miss
Web Applications	Below	Below
Text File Formats	NA	NA
Graphics File Formats	Below	Below
Compressed Archive File Formats	NA	NA
Misnamed Files	NA	NA
Peripheral Memory Cards	NA	NA
Acquisition Consistency	NA	NA
Cleared Devices	NA	Above
Power Loss	Above	Above

Synopsis of TULP 2G

TULP2G is a forensic software framework developed by the Netherlands Forensic Institute (NFI) for extraction and decoding of data stored in electronic devices.[18] "TULP2G is the Dutch acronym for Telefoon Uitlees Programma, 2e Generatie, which roughly translates to Telephone Extraction Program, 2nd Generation.

The TULP2G framework involves an abstract architecture, with distinct plug-in interfaces for data extraction through various means and data decoding of the extracted data, and also a user interface. TULP2G is not designed for presentation, viewing, or searching of extracted information. XML is used as the data storage format, relying on existing tools for these functions. The framework, along with number of different data extraction and decoding plug-ins for cell phones, has been implemented as open source software. The NFI hopes to stimulate efforts in the area of embedded system forensics.

Version 1.1.0.2 of TULP2G can acquire evidence from a phone through different means of communication (i.e., cable, Bluetooth, IrDA) and protocols (i.e., ETSI and Siemens AT commands, IrDA, and OBEX). For GSM phones, it can also acquire SIM data through an external PC/SC reader.

Supported Phones

The tool was designed to work for a wide variety of phones that support one or more common interface standards. Following this approach, the tool performs a logical acquisition using selected commands from the different protocol standards available for USB, IrDA, serial modem interfaces and PC/SC readers. Currently, a modem connection (i.e., serial port, serial over either USB, Bluetooth, or IrDA), a socket connection (i.e., IrDA, Bluetooth), and a PC/SC connection are supported. The appropriate protocol takes place over these connections (i.e., modem:AT_ETSI, AT_Siemens; IrDA: IrMC; Bluetooth: OBEX; PC/SC: SIM chip card data extraction) . The data is then transformed using the corresponding conversion plug-ins (e.g., AT_ETSI, SMS, etc.) and an XML file is produced. A style sheet can be applied to the file to generate reports in a variety of formats and content.

TULP2G can acquire phone calls made, phone calls received, SMS messages, and phone book entries for various phones. It also can acquire more precise data (e.g., sent/received email or calendar, To-Do list etc.), depending on the communication/protocol pair that chosen. TULP2G also acquires IMEI, IMSI and specific data about the phone.

Acquisition Stage

The TULP2G interface provides a series of tabs that follow the logical steps in the extraction procedure, as illustrated in Figure 59. Acquisition can take place through a supported interface. When using a particular connection, an appropriate phone protocol plug-in must also be loaded. For example, a data-link cable acquisition could use the AT-ETSI phone protocol plug-in.

---

[18] Additional information on TULP2G can be found at: **http://tulp2g.sourceforge.net**

78

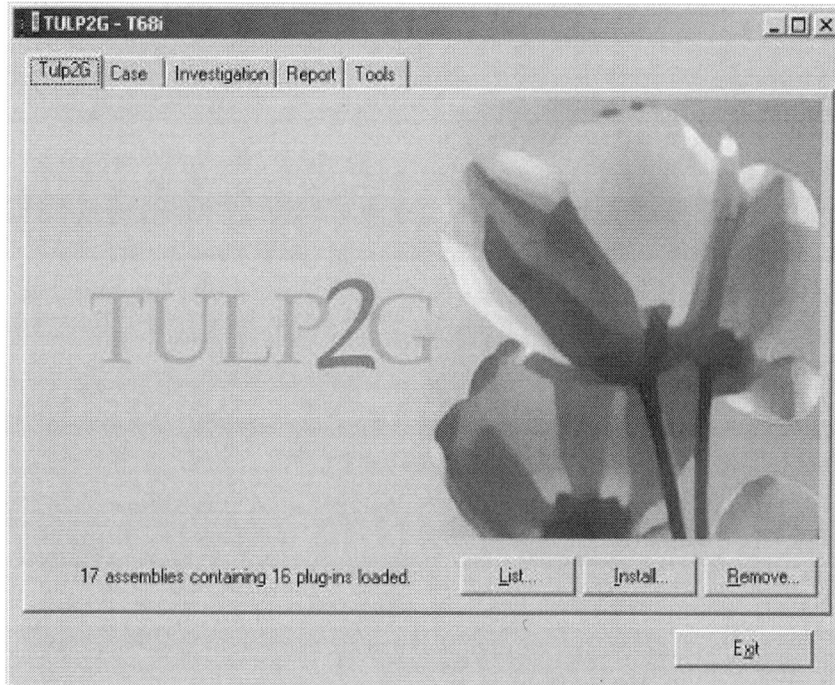

Figure 59: Acquisition Wizard

A case file contains the information about the investigation. Beginning with the "Case" tab, one can create a new case file, load a previous created file, or save an updated case file.

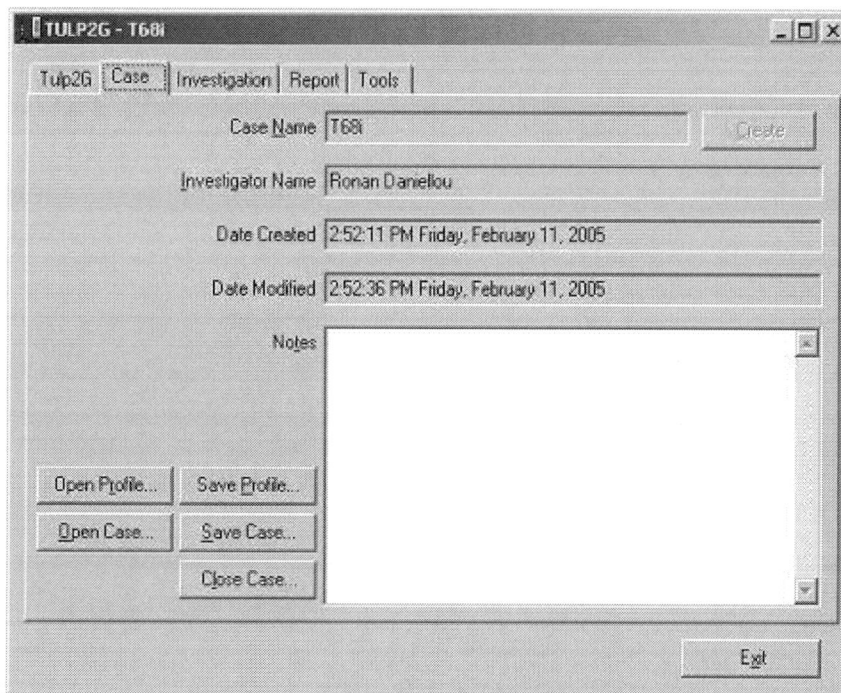

Figure 60: Case Data

An investigation for the case is initiated through the "Investigation" tab by assigning it a name and performing the following steps:

- Choose the communication plug-in: serial (COM port, virtual or not), socket (IrDA and Bluetooth) or PC/SC.
- Pick the protocol plug-in for the tool to use over the communication chosen above (AT_ETSI, AT_SIEMENS, SIM card extraction, OBEX, or IRMC).
- Initiate the acquisition

Note that the examiner can configure port speed, phone and/or SIM commands, test the settings, etc. by clicking on the "Configure" button, at the right of the communication or protocol fields, as illustrated in Figure 61.

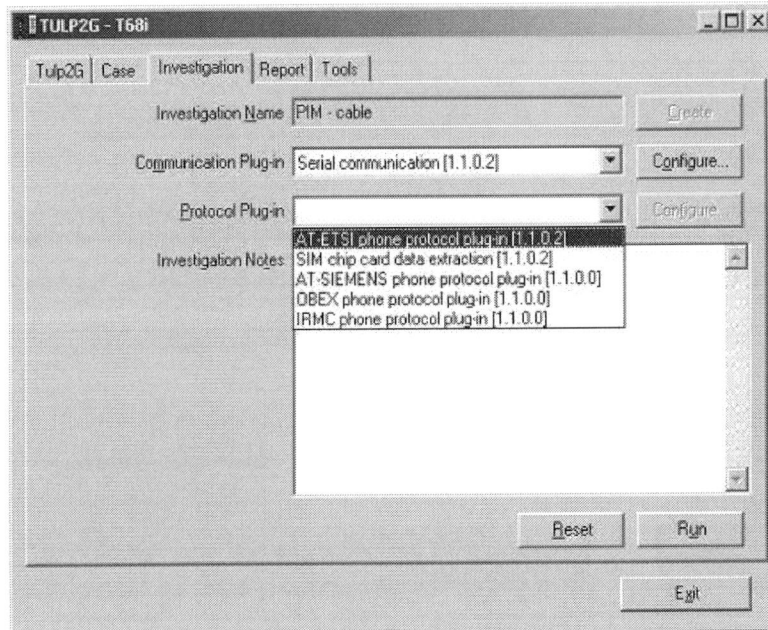

Figure 61: Interface Selection

The progression of the extraction and other details are displayed during acquisition. Figure 62 shows an example screen shot. The details provided may indicate useful information, such as whether certain commands are not supported by the phone.

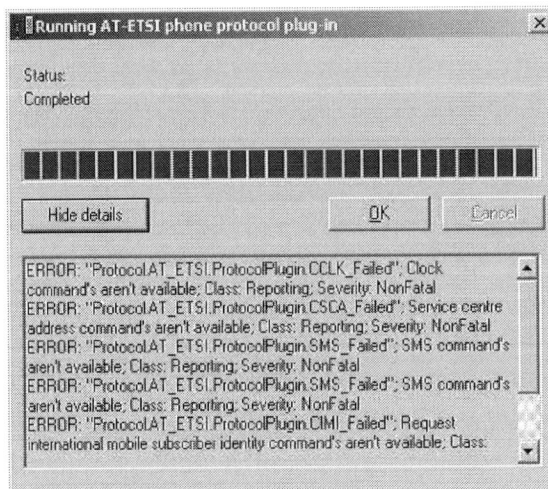

Figure 62: Acquisition Log

80

Search Functionality

Because TULP2G is not a search and analysis tool, after the XML or HTML formatted report is generated, examiners must manually search the report for data, or export the data and use some other search facility.

Graphics Library

Graphic files are stored in the XML output file with <![CDATA[tags and can be converted and recovered using the conversion plug-ins (e.g., OBEX, SMS TPDU). Once the data has been properly converted, individual files can be saved to the desktop for reporting purposes. TULP2G also supports small (16x16) pixel images. Typically, these types of images are embedded in an MMS/EMS message, and are reported alongside the textual content of the SMS/EMS message.

Report Generation

TULP2G generates a set of raw data in XML format, which must be converted to a readable format using the following steps.

- First choose the export plug-in; the default and only choice is XML/HTML. In the configuration panel, choose one of the provided XSL stylesheets.
- Select the conversion plug-in. Several different plug-ins may be required, since each one extracts a distinct set of data.
- Select the investigation(s) from the case to be reported.

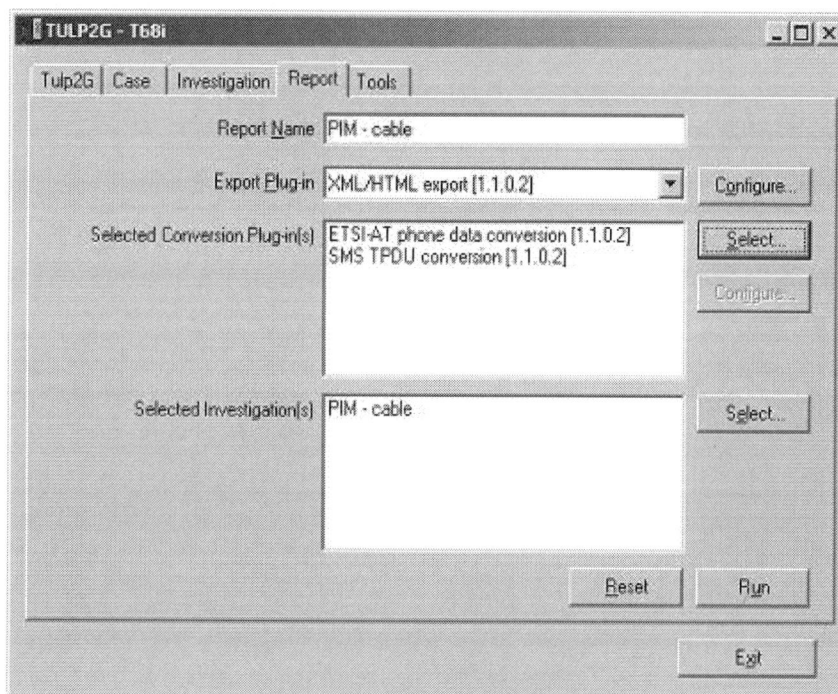

Figure 63: Report Generation

Once generated, the file can be viewed with any appropriate software such as a Web browser or XML editor. Figure 64 shows an excerpt of the generated report.

Investigation *PIM - cable - No entry deleted*

Creation date	2/28/2005 10:12:51 AM
MD5 hash	45FDFFD7C2E25C4B1A88BEEC7530B59A
SHA-1 hash	FC7F362DBE31F026E1DC4BE05A300D3FE718655E

Plug-in info

Plug-in type	Plug-in info
CommunicationPlugin	TULP2G.Communication.SerialPort@TULP2G.Communication.SerialPort, Version=1.1.0.2, Culture=neutral, PublicKeyToken=3480a3624ac48f93
ProtocolPlugin	TULP2G.Protocol.AT_ETSI.ProtocolPlugin@TULP2G.Protocol.AT_ETSI, Version=1.1.0.2, Culture=neutral, PublicKeyToken=3480a3624ac48f93

Selected options
SIM
Phone

Phone

☑ Manufacturer

Manufacturer of the mobile equipment. Typically this item will consist of a single text line containing the name of the manufacturer, but manufacturers may choose to provide more information if desired.

SONY ERICSSON

☑ Request model identification

Specific model of the mobile equipment to which it is connected to. Typically this item will consist of a single text line containing the name of the

Figure 64: Report Excerpt (General Information)

☑ **Request revision identification**

Individual mobile equipment identity. Typically this item will consist of a single text line containing the IMEI, but manufacturers may choose to provide more information if desired.

354545000091955 (International Mobile Equipment Identity(IMEI))

☐ **Request international mobile subscriber identity**

Phonebooks

☐ **ME received calls**

☐ **ME missed numbers (unanswered received)**

☐ **ME phonebook**

☑ **ME dialled calls**

Numbered list of dialled calls stored in the mobile equipment. The number chosen most recently is at the top of the list. When a telephone number already occurs in the SIM or in the GSM telephone with a corresponding description, this description is also given in this list.

☑ **SIM phonebook**

Numbered list of phonebook entries from the phonebook stored in the SIM of the mobile equipment. This list consists of names and telephone numbers, to be entered and changed by the subscriber, which can be chosen easily using the mobile equipment.

Pos.	Name	Number
1	Homer Simpson	9784653210
2	Eric Cartman	1326458790
3	Homer Simpsons	9784653210

Figure 65: Report Excerpt

Scenario Results

Table 19 summarizes the results from applying the scenarios listed at the left of the table to the devices across the top. More information can be found in Appendix H: TULP 2G Results.

Table 19: Results Matrix

Scenario	Device							
	Audiovox 8910	Ericsson T68i	Sony Ericsson P910a	Motorola C333	Motorola V66	Motorola V300	Nokia 6610i	Nokia 6200
Connectivity and Retrieval	Meet	Meet	Meet	Meet	Meet	Meet	Meet	Meet
PIM Applications	Below	Below	Miss	Below	Below	Below	Below	Below
Dialed/Received Phone Calls	Below	Below	Miss	Below	Below	Below	Below	Below
SMS/MMS Messaging	Miss	Below	Miss	Miss	Miss	Below	Below	Below
Internet Messaging	NA	Miss	Miss	NA	NA	Below	NA	NA
Web Applications	Miss	Miss	Miss	NA	NA	Miss	Miss	Miss

Scenario	Device							
	Audiovox 8910	Ericsson T68i	Sony Ericsson P910a	Motorola C333	Motorola V66	Motorola V300	Nokia 6610i	Nokia 6200
Text File Formats	NA	NA	Miss	NA	NA	Miss	Miss	Miss
Graphics File Formats	Miss	Miss	Miss	NA	NA	Miss	Miss	Miss
Compressed Archive File Formats	NA	NA	Miss	NA	NA	NA	Miss	Miss
Misnamed Files	NA	NA	Miss	NA	NA	NA	Miss	Miss
Peripheral Memory Cards	NA	NA	Miss	NA	NA	NA	NA	NA
Acquisition Consistency	NA	NA	NA	NA	NA	NA	NA	NA
Cleared Devices	NA	Meet	Meet	Meet	Meet	Meet	NA	NA
Power Loss	Above	Above	Meet	Above	Above	Above	Above	Above

SIM Card Acquisition

TULP2G version 1.2.0.2 gives examiners the ability to acquire SIM card data using a PC/SC-compatible reader. The acquisition steps followed to acquire data directly from the SIM are the same as acquiring data from a phone, except for selecting PC/SC Chip Card Communication versus a serial or socket. The data fields acquired (e.g., Abbreviated Dialing Numbers, Last Numbers Dialed, Fixed Dialing Numbers, Messages, etc.) are dependent upon the SIM and service provider. The Report facilities operate in a similar fashion as for phone acquisitions, described above.

Table 20 summarizes the results from applying the scenarios listed at the left of the table to the SIMs across the top. More information can be found in Appendix I: TULP 2G – External SIM Results.

Table 20: SIM Card Results Matrix - External Reader

Scenario	SIM		
	5343	8778	1144
Basic Data	Meet	Meet	Meet
Location Data	Below	Below	Below
EMS Data	Meet	Meet	Below
Foreign Language Data	Meet	Meet	Meet

Synopsis of SIMIS

SIMIS version 2.0.13 from Crown Hill is able to acquire information from SIM cards via the PC/SC-compatible card reader that comes with the software.[19] SIMIS provides examiners with a user interface containing a set of tabs providing examiners with the ability to create report notes, import archived reports, search acquired data and PIN administration. SIMIS allows the following data types to be acquired from SIM cards: Abbreviated Dial Numbers (ADN), Fixed Dial Numbers (FDN), International Mobile Subscriber Identity (IMSI), Last Numbers Dialed (LND), Mobile Subscriber Integrated Services Digital Network Number (MSISDN), Short Message Server Parameters (SMSP), SMS Short Messages, Deleted Messages, Public Land Mobile Networks (PLMNS), Forbidden Public Land Mobile Networks (FPLMNS), Location Information, Broadcast Control Channel (BCCH), Cell Broadcast Message Identifier for Data Download (CBMID), Voicemail Number, Integrated Circuit Card Identification (ICCID), Phase ID, Service Provider, Administration Data, Service Dialing Number (SDN) and Capability Configuration Parameters. The tool can also perform a full dump of the card contents for analysis.

Acquisition Stage

The acquisition stage begins by prompting the examiner to select the interface to be used as illustrated below.

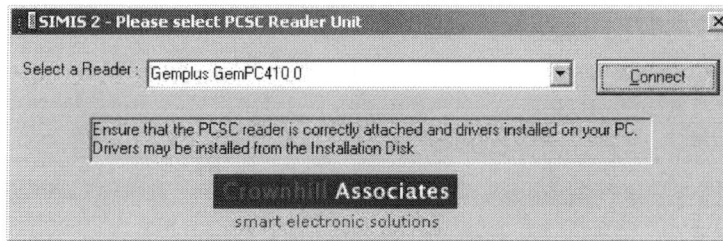

Figure 66: Interface Selection

If the SIM is password protected, the screen below appears. If the examiner does not have the PIN for the SIM, only unprotected data can be acquired from the SIM.

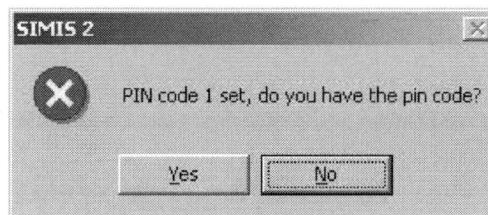

Figure 67: SIM PIN Wizard

If the correct PIN is known, it can be entered into the input window shown in Figure 68 below.

---

[19] Additional information about SIMIS can be found at: http://www.crownhillmobile.com/products.php.

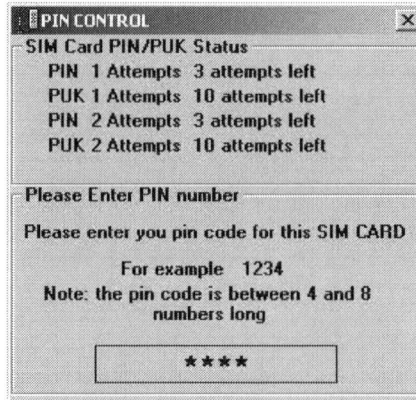

Figure 68: SIM PIN Input

After the correct PIN number has been entered, the examiner can Change, Unlock, Activate, or Deactivate the PIN from the user interface, as illustrated in Figure 69 below.

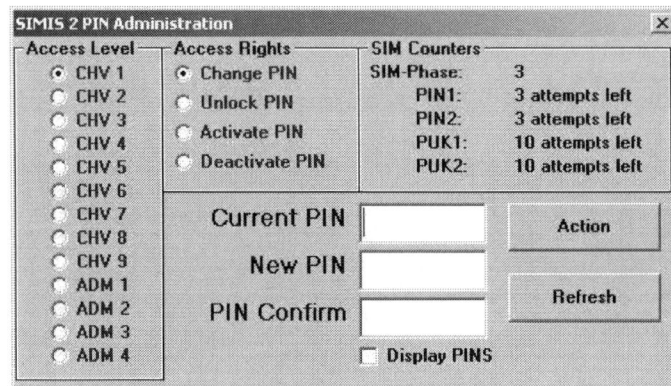

Figure 69: PIN Wizard

The acquisition begins by collecting some information related to the case (i.e., unique file reference, operator name, case number, IMEI, ICCID and Service Provider information found on the SIM) from the examiner that is included in the final report. After a successful acquisition, SIMIS compares the data manually entered (e.g., the ICCID) with the data collected from the SIM for consistency. If a discrepancy is found, the examiner is informed of the inconsistency as illustrated below in Figure 70.

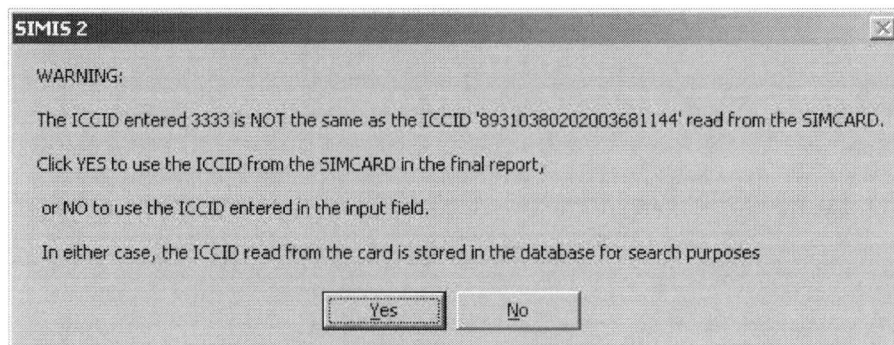

Figure 70: ICCID Warning

86

Figure 71 illustrates a screen shot of the SIMIS table user interface, which allows inspection of the various data fields mentioned above. To begin acquisition the examiner selects the Read SIM button.

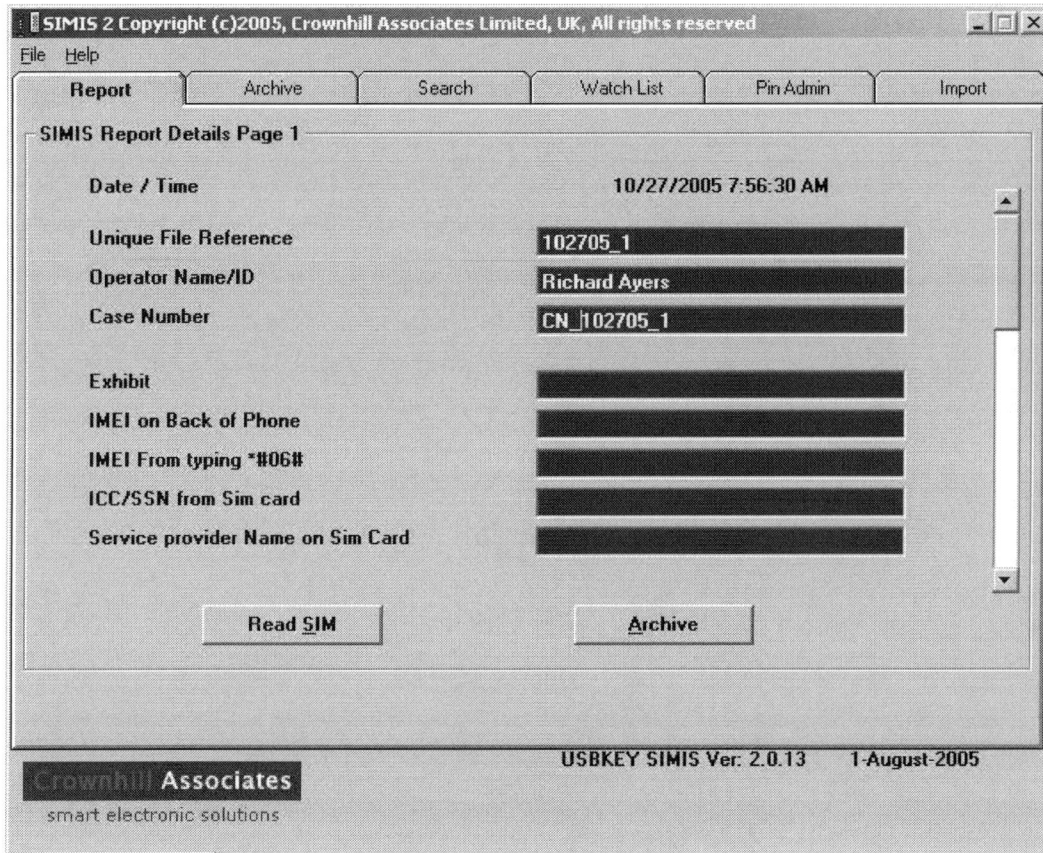

Figure 71: Acquisition Notes

After the data has been acquired, the examiner is presented with an HTML report that is categorized by data item. Figure 72 shows an example screen shot of the menu guided HTML generated report and the ADN entries.

EF_ADN	Case Info		
EF_FDN			
EF_LND	Date /Time:	10/28/2005 8:43:18 AM	
EF_MSISDN	Unique File Reference:	1144_102805_1	
EF_SMS	Operator Name/ID:	richard	
Networks	Case Number:	1144_102805_1	
EF_SDN	Exhibit:		
Card Info	IMEI on Back of Phone:		
Location Info	IMEI From typing *#06#:		
Ext Data	ICC/SSN from SIM card:	ICCID entered: 3333 (Actual Card ICCID: 89310380202003681144)	
VoiceMail	Service provider Name on Sim Card:		
EF_SMSP	Officer in case:		
Deleted SMS	Bag Seal Number:		
Search	Has Bag been resealed:		
EF_KcGPRS	New Bag Seal number if resealed:		
EF_LOCIGPRS	Make of Phone:		
EF_eMLPP	Type of Phone:		
EF_AAeM	Power up display:		
EF_Dck	PIN or PUK Code Entered:		
EF_CPBCCH			
EF_SLL	Notes		
EF_ECCP			
EF_HPLMNwACT			
EF_OPLMNwACT			

Abbreviated Dialled Numbers

```
001: 1234                        Darth
002: 5678                        Vadar
003: 7775551212                  Professor John
004: 0132659842                  暗呵是不中
005: 01133132569632              ClDment SDveil
006:  Blank
007:  Blank
```

Figure 72: Acquisition

After a successful acquisition, SIMIS allows MD5 hashes to be generated ensuring the integrity of the data. MD5 and SHA2 hashes can be created for the SIMIS2 executable, proprietary SIMIS case file, report files, and log files, as illustrated below in Figure 73.

88

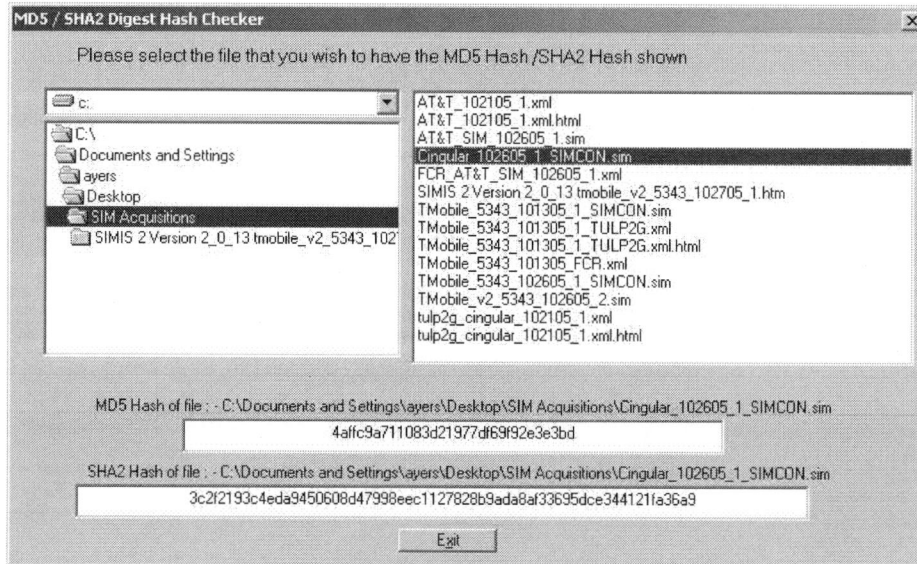

Figure 73: Hash Verification

SIMIS allows examiners to view additional data that is not displayed within the GUI interface. An ASCII dump file (.dmp) can be created in the directory where the SIMIS output data resides by selecting "SIM dump" from the File menu. For example, Forbidden Networks (FPLMNs) can be obtained this way.

Search Functionality

The SIMIS Search engine gives examiners the ability to perform searches on phone numbers contained within a case or the entire database of all cases created. Currently, alpha string searches are not supported. Figure 74 illustrates the interface of the Search tab.

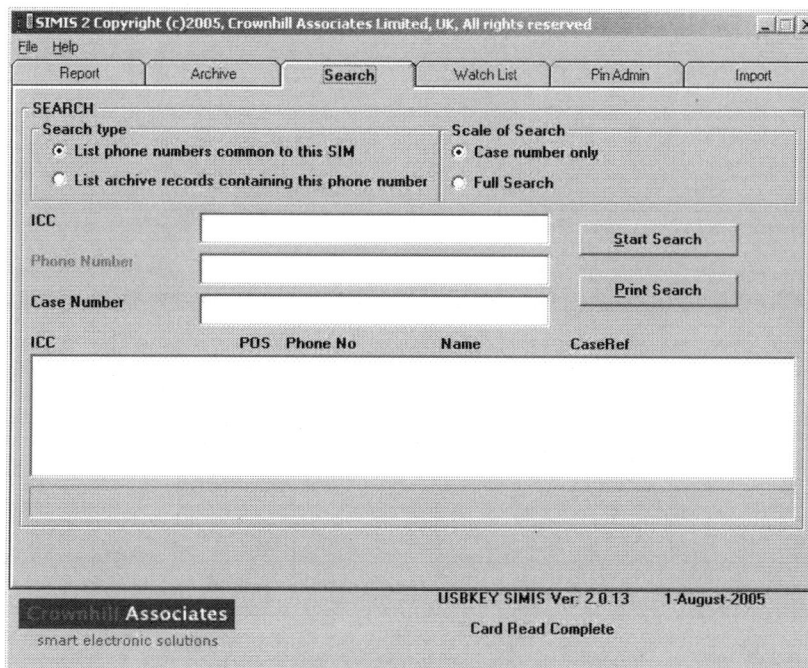

Figure 74: Search Engine

Graphics Library

SIMIS currently does not display graphics files of any type.

Report Generation

The examiner has the opportunity to customize the report with specifics related to the examination before acquisition occurs, which are included in the final version. After selecting the "Read SIM" option and acquiring all data, the report is generated is displayed to the user. The generated report is an HTML-based report that can be viewed with a standard HTML editor. Figure 75 shows an excerpt of the generated report.

```
Short Message Service

Usage

001:     Message received from Network and read
002:     Message received from Network and read
003:     Message received from Network and read
004:     Message received from Network and read
005:     Message received from Network and read
006:     Message received from Network and read
007:     Message has been sent to Network
008:     Message received from Network and read
009:     Message received from Network and read
010:     Message received from Network and read
011:     Message received from Network and read
012:     Free Space

Messages

Message 001

Date Sent             05/10/11 20:08:34 TimeZone GMT+0.00H
Sender                +12404016148
Service Centre        +19703769322
Status                Message received from Network and read

IEI: 00
LENGTH: 00
MESSAGE REF: 00
STATUS: 01
SC Address length: 07
SC Address type: 91
   Type of number: International
   Numbering plan identifier: E.164
SC Address: 19703769322
Message Type Indicator: 00
Message Type: SMS-DELIVER / SMS-DELIVER REPORT
More Messages To Send: Yes
Status Report Indication: No
Reply Path: No
Originating Address Length: 0B
Originating Address type: 91
   Type of number: International
   Numbering plan identifier: E.164
Originating Address: 12404016148
Protocol Identifier: Default
Data Coding Scheme: GSM Default Alphabet
SC Timestamp: 50011102804300
   decoded: 05/10/11 20:08:34
Time Zone: GMT+0.00H
User Data Length: 13
   decimal: 19
Message: Active incoming sms
```

Figure 75: Report Excerpt

90

Scenario Results

Table 21 summarizes the results from applying the scenarios listed at the left of the table to the SIMs across the top. More information can be found in Appendix J: SIMIS Results.

Table 21: Results Matrix

Scenario	SIM		
	5343	8778	1144
Basic Data	Meet	Meet	Meet
Location Data	Meet	Below	Below
EMS Data	Below	Below	Below
Foreign Language Data	Below	Below	Below

Synopsis of ForensicSIM

The ForensicSIM toolkit from Radio Tactics is able to acquire information from SIM cards via a PC/SC-compatible card reader that comes with the software. Before the data contents of the SIM can be analyzed with the Forensic Analysis software, the examiner must use the ForensicSIM acquisition terminal, a stand-alone device, to create a copy of the target SIM on a separate storage card. To copy the SIM, the examiner must log on to the acquisition terminal with a username and PIN number. The acquisition terminal then walks the examiner through the process of entering the target SIM and creating duplicate copies on the provided blank SIM cards (i.e., a Master SIM copy used for storage in case of an evidence dispute, a Prosecution SIM copy that serves as a working duplicate for evidence recovery and analysis, and a Defense SIM copy that serves as a working duplicate issued to the defense). In addition to the aforementioned SIM copy cards, Radio Tactics provides the option to create an Access Card. The Access Card holds a copy of data from the target SIM and permits the target handset to be examined without the risk of connecting to the cell network.

Once the target SIM has been successfully duplicated, the ForensicSIM software along with the SIM reader can be used to create a report of the data found on a duplicate working SIM. The ForensicSIM toolkit allows the following data types to be recovered: Subscriber/User related files, Phone Number related files, SMS related files, Network related files and General SIM information. Each individual field contains more specific meta-data about each type.

Acquisition Stage

The acquisition process begins by guiding the examiner through a succession of screens about the examination. The examiner has the option of generating a report from either the SIM card or a saved case. As mentioned earlier, the ForensicSIM PC/SC reader is used along with the Radio Tactics ForensicSIM analysis software to acquire data from the SIM card. The initial screen for report and case generation is illustrated below in Figure 76.

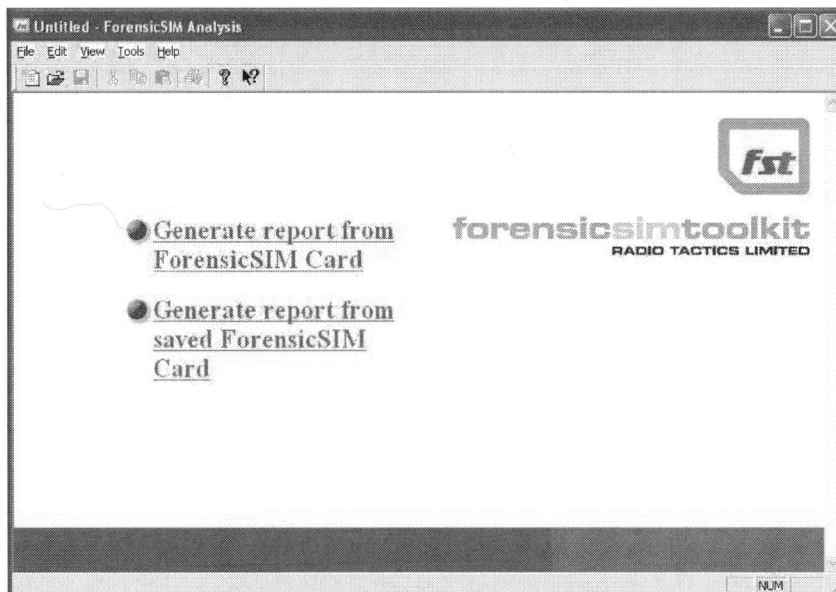

Figure 76: Acquisition Wizard

The examiner has the option of creating a standard report or an advanced report. A standard report displays only card identification information, phonebook, and SMS text messages. The advanced report displays additional information recovered from the SIM card. After the data on the SIM has been successfully acquired, the examiner is then asked to enter case specific information (i.e., Operator, Operator Name, Date/Time, Reference No., Case Reference No., Case Officer, Exhibit Reference, Exhibit Seal No., Exhibit Reseal No., Phone Make, Phone Type, IMEI and PIN/PUK codes, if known). The information entered by the examiner is contained in the finalized report. Illustrated below in Figure 77 is a snapshot of the User Interface after an acquisition has completed. The user interface provides a tree structure where specific data items can be selected for analysis.

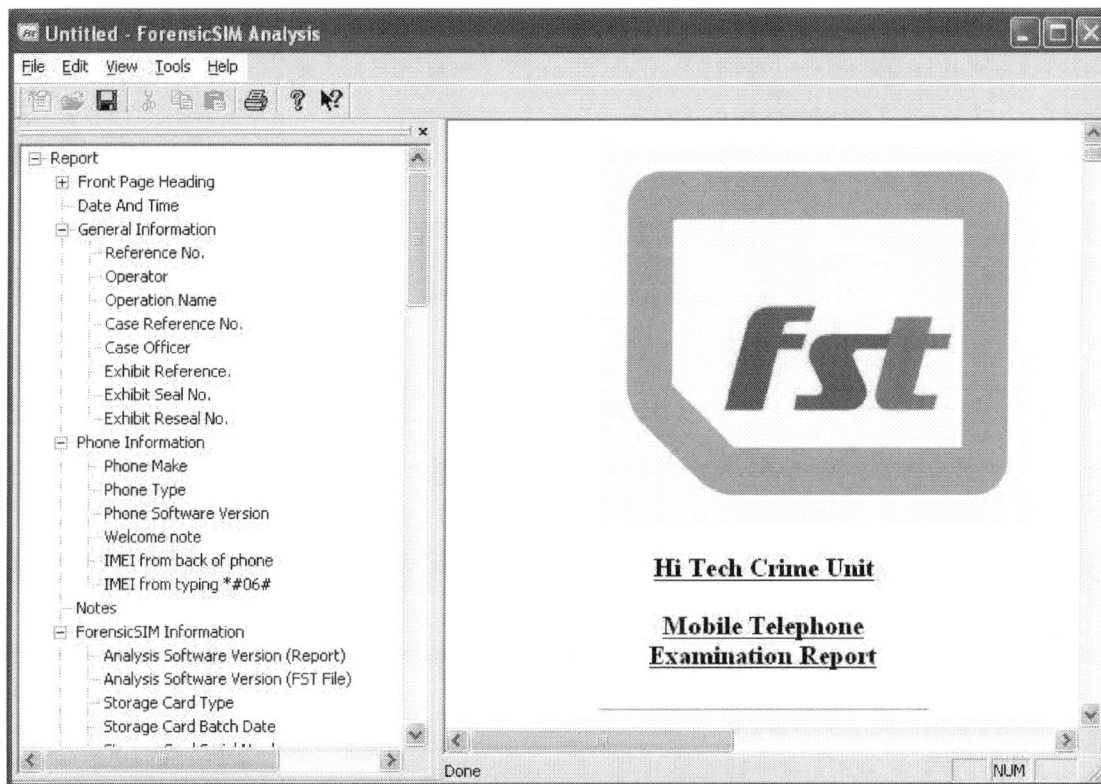

Figure 77: User Interface

Search Functionality

ForensicSIM does not have a search facility. After the report is generated, examiners can manually search the data it contains or use an automated search engine appropriate for the exported file type of the report.

Graphics Library

ForensicSIM does not display graphics files of any type.

Report Generation

After a successful acquisition has taken place the examiner then has the opportunity to create a customized report of findings that pertain to the case. To finalize the report, the export option is selected, which allows specific data fields to be incorporated into the report. Figure 78 gives a snapshot of the Report Wizard that allows examiners to select the data fields for the finalized report.

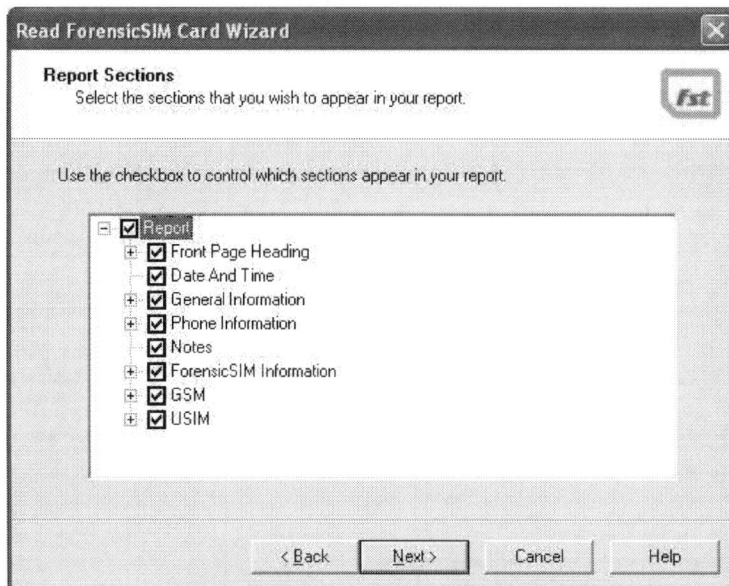

Figure 78: Report Wizard

The examiner has the choice of exporting the final report in the formats illustrated below in Figure 79.

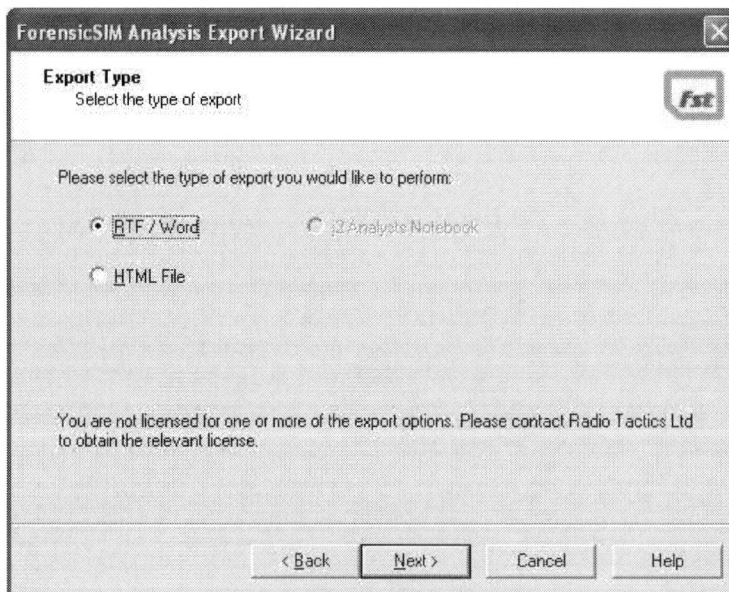

Figure 79: Report Output Type

Figure 80 and Figure 81 are snapshots of the final generated report.

General Information

Reference No.	032805_1
Operator	rpa
Operation Name	richard ayers
Case Reference No.	crn_032805_1
Case Officer	rpa_1
Exhibit Reference.	er_032805_1
Exhibit Seal No.	es_032805_1
Exhibit Reseal No.	ern_032805_1

Phone Information

Phone Make	Nokia
Phone Type	6220
Phone Software Version	Nokia OS

ForensicSIM Information

Analysis Software Version (Report)	1.2.5.0
Analysis Software Version (FST File)	Not present
Storage Card Type	Defence
Storage Card Batch Date	21/06/04
Storage Card Serial Number	0000001732
Storage Card Version	1
Control Card Id	3B6600FF4A434F503130FFFFFFFFFFFFFF020065
Control Card User Id	0000
Acquisition Hardware Type	Desktop (MCT5000)
Acquisition Software Version	1.2.1.0
Card Write Start Time	28/03/05 19:39.59
Card Write End Time	Uncalibrated
Number of bytes read	14025
Number of bytes copied	14025
Hash Type	MD5
Hash Code	FCD9A3BBAEF3639C8898C70BED578418

Mobile Station Numbers (MSISDN)

Nr	Name	Telephone Number	Type of number	Numbering Plan	Extension Id
1	My Mobile #	1240731xxxx	International	ISDN/Telephony	None

International Mobile Subscriber Identity (IMSI)

IMSI
310380042199423 MCC: 0310, MNC: 38

Service Provider Name (SPN)

Provider Display Status	Provider Name
Not Displayed	AT&T Wireless

Figure 80: Report Excerpt

95

Operator Name String (ONS)

PLMN name
AT&T Wireless

Abbreviated Dialled Numbers

Nr	Name	Telephone Number	Type of number	Numbering Plan	Extension Id
1	homer simpson	9784653210	None	ISDN/Telephony	None

Last Number Dialled

Nr	Name	Telephone Number	Type of number	Numbering Plan	Extension Id
1		9784653210	None	ISDN/Telephony	None
3		301972xxxx	None	ISDN/Telephony	None
4		301975xxxx	None	ISDN/Telephony	None

Mailbox Numbers

Nr	Name	Number	Type of number	Numbering Plan	Extension Id
1	Voice Mail	1443280xxxx	International	ISDN/Telephony	None

SIM Card SMS Memory Usage

Read messages

There are 2 read messages.

SMS Message	
Message Location	1
Status	Read
Message Type	SMS-DELIVER
Originating Address	+12404016148
Service Centre Address	+19703769328
Service Centre Timestamp	25-02-05 10:41:15 GMT -5:00
Data Coding Scheme	Default alphabet
User Data Header	Not present
User Data	This is to determine if sms messages can be properly acquired?

SMS Message	
Message Location	2
Status	Read
Message Type	SMS-DELIVER
Originating Address	12407310023
Service Centre Address	+19703769328
Service Centre Timestamp	09-03-05 16:42:38 GMT +0:00
Data Coding Scheme	Default alphabet
User Data Header	Not present
User Data	ATT Wireless: You received a picture message your phone can't display. See it at the URL below-expires soon. Use code 4jbvpc2k http://www.attwireless.com/inbox

Figure 81: Report Excerpt

Scenario Results

Table 22 summarizes the results from applying the scenarios listed at the left of the table to the SIMs across the top. More information can be found in Appendix K: ForensicSIM Results.

Table 22: Results Matrix

Scenario	SIM		
	5343	8778	1144
Basic Data	Meet	Miss	Below
Location Data	Meet	Miss	Below
EMS Data	Below	Miss	Below
Foreign Language Data	Below	Miss	Below

Synopsis of Forensic Card Reader

Forensic Card Reader (FCR)[20] version 1.01 is able to acquire information from SIM cards via the PC/SC Chipy reader. The FCR PC/SC USB card reader and FCR software give examiners the ability to capture data such as the ICC ID, IMSI, incoming/outgoing calls, abbreviated call numbers, SMS messages and location data.

Acquisition Stage

After installing the FCR software, connecting the USB reader, and selecting the proper PC/SC reader, the SIM data content acquisition begins by clicking the Read button on the beginning wizard screen as illustrated in Figure 82.

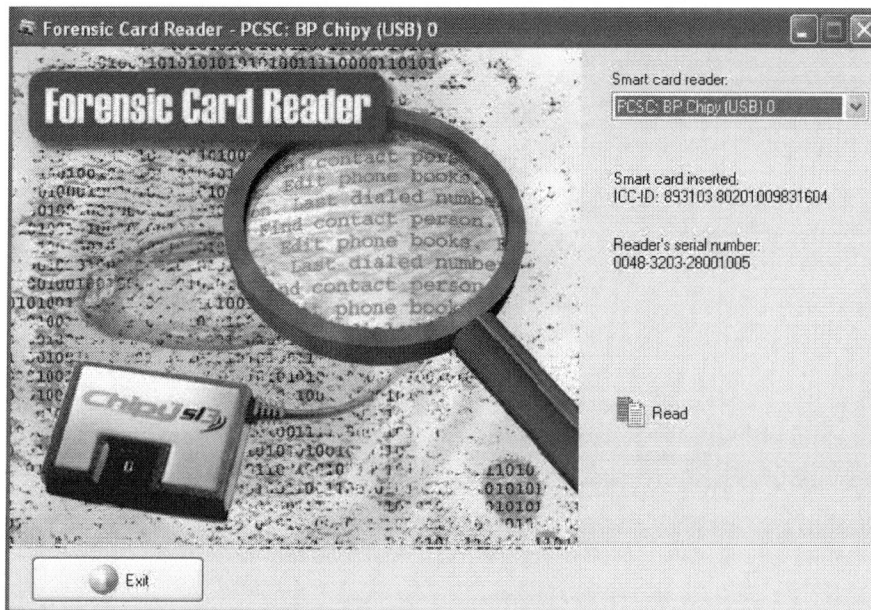

Figure 82: Interface Selection

The wizard allows the examiner to select data elements to be acquired from the SIM and specify the output directory, as illustrated below in Figure 83.

---

[20] Additional information on FCR can be found at: **http://www.becker-partner.de/forensic/intro_e.htm**

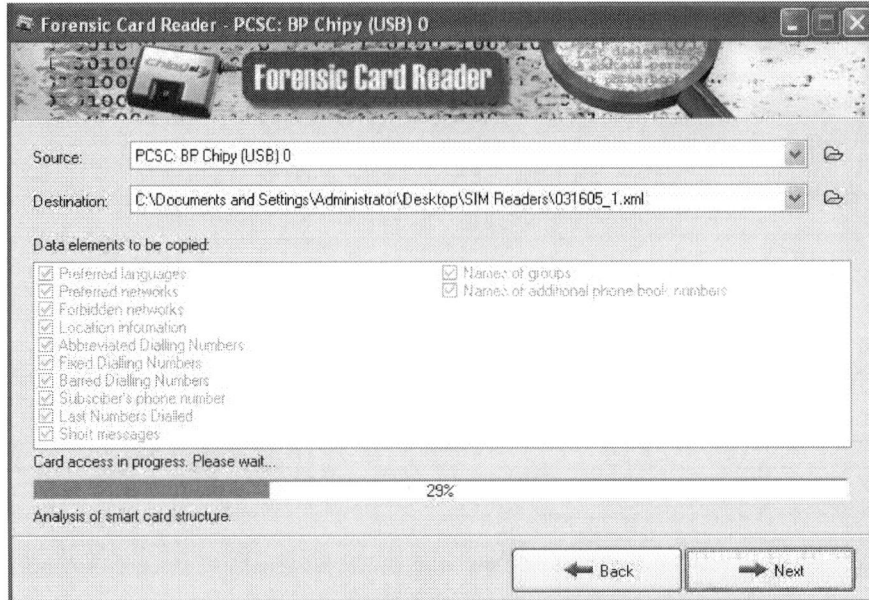

Figure 83: Acquisition Wizard

The acquisition is finalized by clicking the Read button, as illustrated below in Figure 84. Data elements and acquisition progress are displayed, allowing the progress of the acquisition to be monitored.

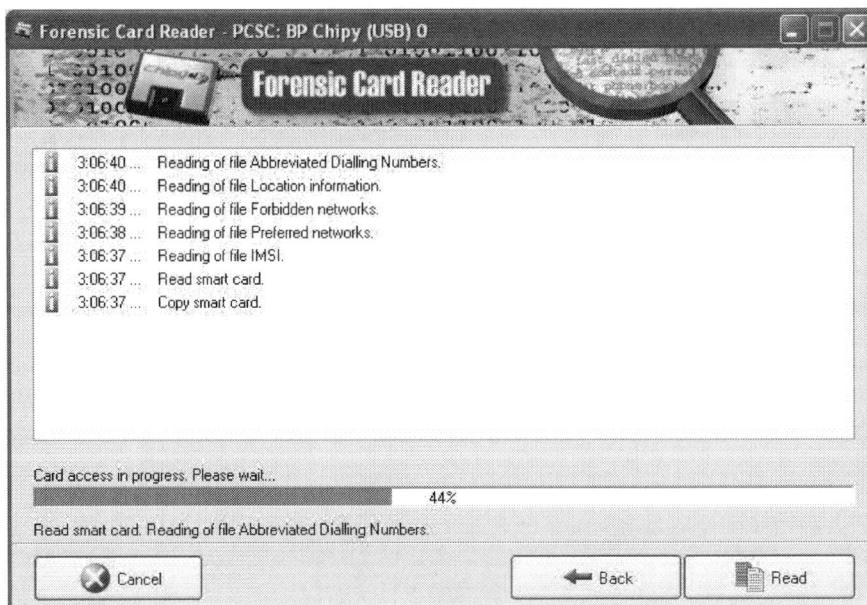

Figure 84: SIM Acquisition

Search Functionality

FCR does not have a built-in search engine or search facilities. Once the XML-formatted report is generated, examiners can manually search the data it contains or use an appropriate search tool.

Graphics Library

FCR does not display graphics files of any type. The FCR application only allows examiners to export textual data in the XML format.

Report Generation

Once the finalized report is generated, the file can be viewed with an appropriate Web browser or XML editor.

```xml
<?xml version="1.0" encoding="UTF-8" standalone="yes" ?>
<scard version="1.1" reader="0048-3203-28001005">
  <icc_id>893103 80201009831604</icc_id>
  <imsi>9310380042199423</imsi>
  <phone_book type="ADN">
    <entry>
      <name>homer simpson</name>
      <number>9784653210</number>
    </entry>
  </phone_book>
  <phone_book type="MSISDN">
    <entry>
      <name>My Mobile #</name>
      <number>+12407310478</number>
    </entry>
  </phone_book>
  <phone_book type="LND">
    <entry>
      <number>9784653210</number>
    </entry>
    <entry>
      <number>9784653210</number>
    </entry>
    <entry>
      <number>+14432803092</number>
    </entry>
  </phone_book>
  <messages>
    <message type="sms" processed="True">
      <time>2/25/2005 10:41:15 AM</time>
      <smsc>+19703769328</smsc>
      <orig>+12404016148</orig>
      <text>This is to determine if sms messages can be properly acquired?</text>
    </message>
    <message type="sms" processed="True">
      <time>3/9/2005 4:42:38 PM</time>
      <smsc>+19703769328</smsc>
      <orig>12407310023</orig>
      <text>ATT Wireless: You received a picture message your phone can't display. See it at the URL
        http://www.attwireless.com/inbox</text>
    </message>
  </messages>
```

Figure 85: Report

Scenario Results

Table 23 summarizes the results from applying the scenarios listed at the left of the table to the SIMs across the top. More information can be found in Appendix L: Forensic Card Reader Results.

Table 23: Results Matrix

Scenario	SIM		
	5343	**8778**	**1144**
Basic Data	Below	Below	Below
Location Data	Below	Below	Below
EMS Data	Below	Below	Below
Foreign Language Data	Below	Below	Below

Synopsis of SIMCon

SIMCon version 1.1 can acquire information from SIM cards via a PC/SC-compatible reader. The SIMCon software gives examiners the ability to capture and examine data such as the Card Identity (ICCID), Stored Dialing Numbers (ADN), Fixed Dialing numbers (FDN), Subscriber Number (MSISDN), Last Numbers Dialed (LND), SMS Messages, Subscriber Identity (IMSI), Ciphering Key (Kc) and Location Information (LOCI).

Acquisition Stage

Once proper connectivity is established with the SIM, the examiner is prompted for the correct PIN, if the SIM is protected, before acquisition begins. If the SIM does not contain a PIN acquisition begins by clicking OK as illustrated below in Figure 86.

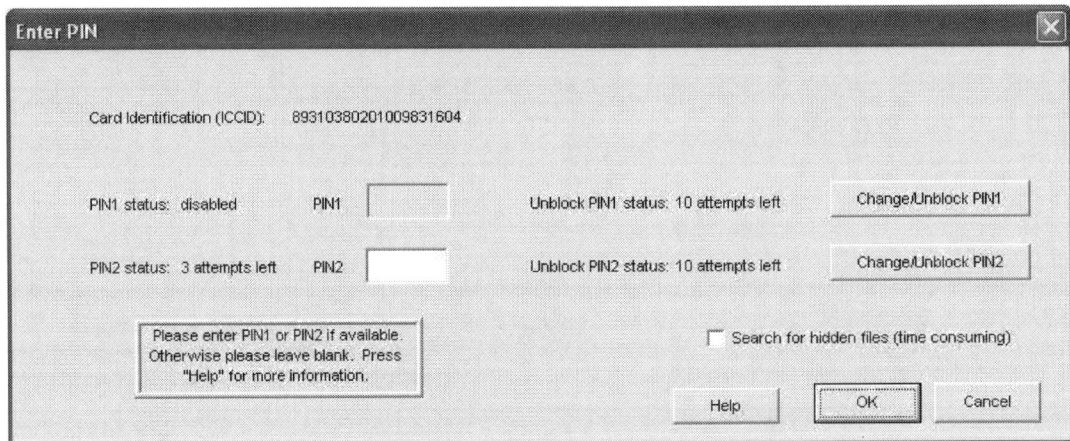

Figure 86: Acquisition Wizard

After a successful acquisition, the entire SIM contents can be saved and stored in the SIMCon .sim proprietary format for later processing. SIMCon allows for examiners to search for hidden files by checking the "Search for hidden files" checkbox, which potentially may uncover pieces of valuable data relevant to the case. SIMCon uses an internal hashing facility to ensure the integrity of cases and detect whether tampering occurred during storage. SIMCon uses the SHA1 algorithm to compute a hash for each file as it is read from the card. Selecting "Verify Hash" in the "File" menu causes SIMCon to recompute all hashes and check that the original file is consistent with the reopened case. Figure 87 provides a snapshot of the User Interface. The display is divided into three primary panes consisting of a tree structure of data fields, individual data within a selected field, and a textual or hexadecimal view of the data selected for investigation.

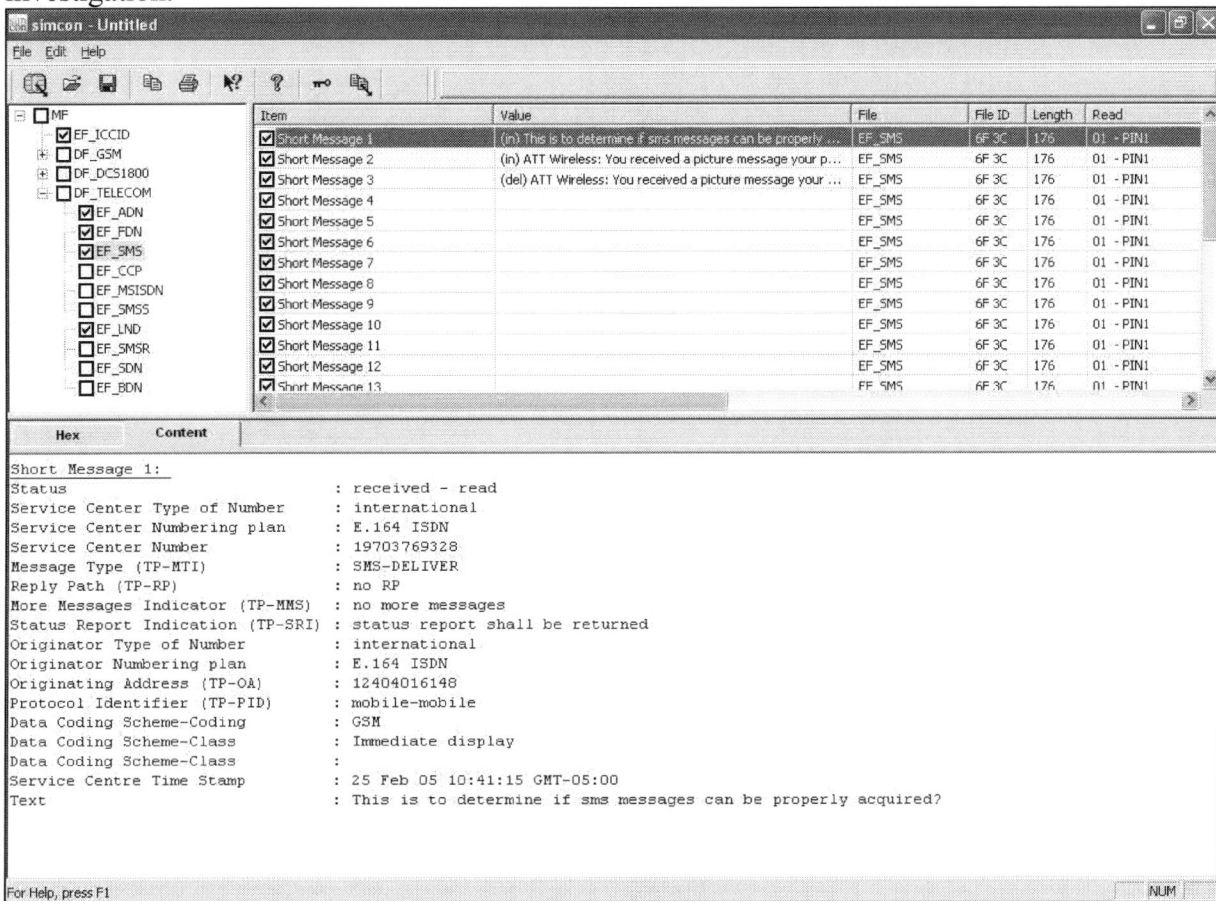

Figure 87: User Interface

Search Functionality

SIMCon does not have a built-in search engine or search facilities. However, data can be viewed manually by browsing through the tree structure and selecting individual data items. A textual and hexadecimal representation is provided in the lower pane. Also, once the formatted report is generated, examiners can manually search the data it contains or use an appropriate automated search tool.

Graphics Library

SIMCon only supports small (16x16) and large (32x32) pixel images. Typically, these types of images are embedded in an MMS/EMS message, and are reported alongside the textual content of the SMS/EMS message. SIMCon does not report or display any other graphics files, such as, graphic files embedded in MMS messages.

Report Generation

Report generation begins by prompting the examiner with case specific details such as investigator name, date/time, case id, evidence number, and notes specific to the investigation as illustrated below in Figure 88.

101

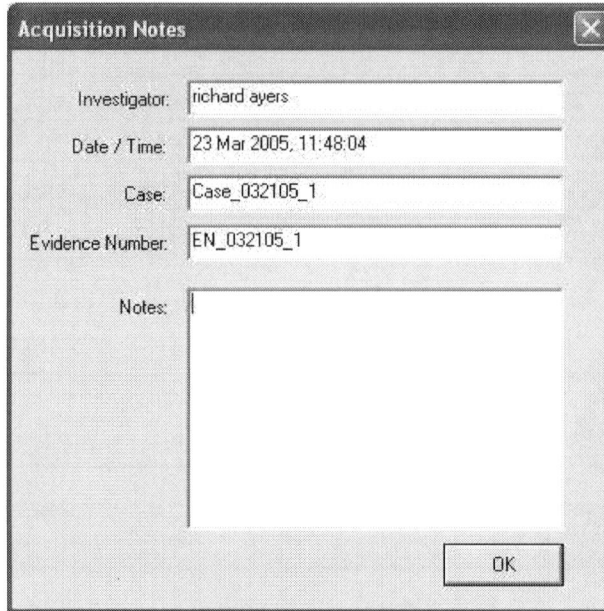

Figure 88: Acquisition Notes

The final report can either be sent to a printer or saved to a file. As illustrated below in Figure 89, the examiner can include checked data items, highlighted items, or all items in the final report.

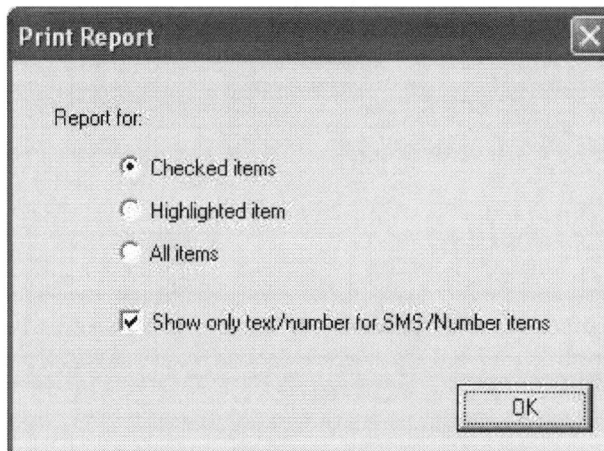

Figure 89: Report Generation

An excerpt of the report is illustrated below in Figure 90. The report includes information relevant to the case as mentioned earlier and data items selected by the examiner that are relative to the case or incident.

```
Investigator      : richard ayers
Case              : Case_032105_1
Evidence number   : EN_032105_1
Timestamp         : 23 Mar 2005, 11:48:04
Notes             :

Card Identity: 8931038020100983I604

International Mobile Subscriber Identity (IMSI): 310380042199423 - USA 3650 AT&T, IMEI: 351102/50/2/******

International Mobile Subscriber Identity (IMSI): 310380042199423 - USA 3650 AT&T, IMEI: 351102/50/2/******

Abbreviated Dialling Number 1:
Identifier                          : homer simpson
Type of Number                      :
Numbering plan                      : E.164 ISDN
Number                              : 9784653210
Cabability config id               : FF


Short Message 1:
Status                              : received - read
Service Center Type of Number       : international
Service Center Numbering plan       : E.164 ISDN
Service Center Number               : 19703769328
Message Type (TP-MTI)              : SMS-DELIVER
Reply Path (TP-RP)                 : no RP
More Messages Indicator (TP-MMS)   : no more messages
Status Report Indication (TP-SRI)  : status report shall be returned
Originator Type of Number          : international
Originator Numbering plan          : E.164 ISDN
Originating Address (TP-OA)        : 1240401614B
Protocol Identifier (TP-PID)       : mobile-mobile
Data Coding Scheme-Coding           : GSM
Data Coding Scheme-Class            : Immediate display
Data Coding Scheme-Class            :
Service Centre Time Stamp           : 25 Feb 05 10:41:15 GMT-05:00
Text                               : This is to determine if sms messages can be properly acquire

Last number dialled 1:
Identifier                          :
Type of Number                      :
Numbering plan                      : E.164 ISDN
Number                              : 9784653210
cabability config id               : FF
```

Figure 90: Report Excerpt

Scenario Results

Table 24 summarizes the results from applying the scenarios listed at the left of the table to the SIMs across the top. More information can be found in Appendix M: SIMCon Results.

Table 24: Results Matrix

Scenario	SIM		
	5343	8778	1144
Basic Data	Meet	Below	Below
Location Data	Meet	Meet	Meet
EMS Data	Meet	Meet	Meet
Foreign Language Data	Meet	Meet	Meet

103

Conclusions

Forensic examination of cellular devices is a growing subject area in computer forensics. Consequentially, cell phone forensic tools are a relatively recent development and in the early stages of maturity. Forensic examination tools translate data to a format and structure that is understandable by the examiner and can be effectively used to identify and recover evidence. However, tools may contain some degree of inaccuracies. For example, the tool's implementation may contain a programming error; a specification used by the tool to translate encoded bits into data comprehensible by the examiner may be inaccurate or out of date; or the protocol structure generated by the cellular device as input may be incorrect, causing the tool to function improperly. In addition, a knowledgeable suspect may tamper with device information to foil the workings of a tool or apply a wiping tool to remove or eliminate data. Over time, experience with a tool provides an understanding of its limitations, allowing an examiner to compensate where possible for any shortcomings or to turn to other means of recovery.

While the tools discussed in this paper generally performed well and have adequate functionality, new versions are expected to improve and better meet investigative requirements. For instance, during the course of preparing this report, a new version for nearly every tool was issued, which included support for additional phones and enhanced functionality.

The following criteria highlight some items to consider when choosing among available tools:
- Usability – the ability to present data in a form that is useful to an investigator.
- Comprehensive – the ability to present all data to an investigator so that evidence pertaining to an investigation can be identified.
- Accuracy – the quality that the output of the tool has been verified and a margin of error ascertained.
- Deterministic – the ability for the tool to produce the same output when given the same set of instructions and input data.
- Verifiable – the ability to ensure accuracy of the output by having access to intermediate translation and presentation results.
- Acceptance – the degree of peer review and agreement about the methodology or technique used by the tool.
- Quality – the technical support, reliability, and maintenance provided by the manufacturer
- Capability – the supported devices, feature set, performance, and richness of features with regard to flexibility and customization
- Affordability – the cost versus the associated benefits in productivity

Glossary of Acronyms

ADN (Abbreviated Dialing Numbers) – phone book entries kept on the SIM.

CDMA (Code Division Multiple Access) – a spread spectrum technology for cellular networks based on the Interim Standard-95 (IS-95) from the Telecommunications Industry Association (TIA).

EDGE (Enhanced Data for GSM Evolution) – an upgrade to GPRS to provide higher data rates by joining multiple time slots.

EMS (Enhanced Messaging Service) – an improved message system for GSM mobile phones allowing picture, sound, animation and text elements to be conveyed through one or more concatenated SMS messages.

ESN (Electronic Serial Number) – a unique 32-bit number programmed into CDMA phones when they are manufactured.

FCC ID (Federal Communications Commission identification number) – an identifier found on all wireless phones legally sold in the US, which is issued by the FCC.

FDN (Fixed Dialing Numbers) – a set of phone numbers kept on the SIM that the phone can call exclusively of any others (i.e., all other numbers are disallowed).

FPLMN (Forbidden PLMNs) – a list of Public Land Mobile Networks (PLMNs) maintained on the SIM that the phone cannot automatically contact, usually because service was declined by a foreign provider.

GID1 (Group Identifier Level 1) – an identifier for a particular SIM and handset association, which can be used to identify a group of SIMs involved in a particular application.

GID2 (Group Identifier Level 2) – a GID1-like identifier.

GPRS (General Packet Radio Service) – a packet switching enhancement to GSM and TDMA wireless networks to increase data transmission speeds.

GRPSLOCI (GPRS Location Information) – the Routing Area Information (RAI), Routing Area update status, and other location information maintained on the SIM.

GSM (Global System for Mobile Communications) – a set of standards for second generation, cellular networks currently maintained by the 3rd Generation Partnership Project (3GPP).

HTTP (HyperText Transfer Protocol) – a standard method for communication between clients and Web servers.

ICCID (Integrated Circuit Card Identification) – a unique and immutable identifier maintained within the SIM.

iDEN (Integrated Digital Enhanced Network) – a proprietary mobile communications technology developed by Motorola that combine the capabilities of a digital cellular telephone with two-way radio.

IM (Instant Messaging) – a facility for exchanging messages in real-time with other people over the Internet and tracking the progress of the conversation.

IMEI (International Mobile Equipment Identity) – a unique number programmed into GSM and UMTS mobile phones.

IMSI (International Mobile Subscriber Identity) – a unique number associated with every GSM mobile phone user.

IMAP (Internet Message Access Protocol) – a method of communication used to read electronic messages stored in a remote server.

LND (Last Numbers Dialed) – a log of last numbers dialed, similar to that kept on the phone, but kept on the SIM without a timestamp.

LOCI (Location Information) – the Location Area Identifier (LAI) of the phone's current location, continuously maintained on the SIM when the phone is active and saved whenever the phone is turned off.

MMS (Multimedia Messaging Service) – an accepted standard for messaging that lets users send and receive messages formatted with text, graphics, photographs, audio, and video clips.

MSISDN (Mobile Subscriber Integrated Services Digital Network) – the international telephone number assigned to a cellular subscriber.

PIM (Personal Information Management) – data types such as contacts, calendar entries, tasks, notes, memos and email that may be synchronized from PC to device and vice-versa.

POP (Post Office Protocol) – a standard protocol used to receive electronic mail from a server.

SIM (Subscriber Identity Module) – a smart card chip specialized for use in GSM equipment.

SMS (Short Message Service) – a mobile phone network facility that allows users to send and receive alphanumeric text messages of up to 160 characters on their cell phone or other handheld device

SMS (Short Message Service) Chat – a facility for exchanging messages between mobile phone users in real-time via SMS text messaging, which allows previous messages from the same conversation to be viewed.

SMTP (Simple Mail Transfer Protocol) – the primary protocol used to transfer electronic mail messages on the Internet.

UMTS (Universal Mobile Telecommunications System) – a third-generation (3G) mobile phone technologies standardized by the 3GPP as the successor to GSM.

USIM (UMTS Subscriber Identity Module) – a module similar to the SIM in GSM/GPRS networks, but with additional capabilities suited to 3G networks.

WAP (Wireless Application Protocol) – a standard that defines the way in which Internet communications and other advanced services are provided on wireless mobile devices.

WIM (WAP identity module) – a security module implemented in the SIM that provides a trusted environment for using WAP related applications and services on a mobile device via a WAP gateway.

WiFi (Wireless Fidelity) – a generic term that refers to a wireless local area network that observes the IEEE 802.11 protocol.

WML (Wireless Markup Language) – a stripped down version of HTML to allow mobile devices to access Web sites and pages that have been converted from HTML to the more basic text format supported.

XHTML (eXtensible HyperText Markup Language) – a unifying standard that brings the XML benefits of easy validation and troubleshooting to HTML.

XML (Extensible Markup Language) – a flexible text format designed to describe data for electronic publishing.

Appendix A: PDA Seizure Results

The scenarios were performed on a Forensic Recovery of Evidence Device (FRED) running Windows XP SP2. PDA Seizure versions 3.0.1.26 to 3.0.3.89 were used to acquire data from Palm OS, Pocket PC and BlackBerry devices with cell phone capabilities.

Blackberry 7750

The following scenarios were executed on a Verizon CDMA BlackBerry 7750 Java-based wireless handheld device running v3.7.1.36 (Platform 1.4.0.37). PDA Seizure version 3.0.1.35 was used for acquisition.

Connectivity and Retrieval: The device contents were successfully acquired. If authentication mechanisms are applied, the proper pass-phrase must be provided within 10 attempts. The correct pass-phrase must be provided for both Memory and Database acquisitions. Basic subscriber and service provider information (e.g., ESN, MSID, FCCID) were not found. The reported memory size is inconsistent with the total size found on the device via the Options->Status screen. (Meet)

PIM Applications: All PIM data (i.e., Address Book, Calendar, Tasks, Memos) was found in their corresponding databases (i.e., Address Book, Calendar, Tasks, Memos database) and reported. All deleted PIM data was found and reported in the memory window. (Meet)

Dialed/Received Phone Calls: All dialed/received phone calls were found and reported in the Phone Call Log, the Phone Hotlist database, and the memory window when conducting a search. Deleted calls were found in the memory window. (Meet)

SMS/MMS Messaging: All active incoming and outgoing SMS messages were found and reported in the Messages folder and the memory window. All deleted SMS messages were found and reported in the memory window. MMS messages are not supported. (Meet)

Internet Messaging: All data content associated with sent and received email messages was found and reported in the Messages database and the memory window. Deleted messages were found and reported in the memory window. The Blackberry device does not support Instant Messaging; therefore, this part of the scenario does not apply. (Meet)

Web Applications: N.A. – In order for the Verizon BlackBerry 7750 to utilize the Web viewer, Verizon customers must purchase a third party solution such as MobileWeb4U Mobile Web WAP Gateway. Verizon does not provide a public WAP gateway with this capability. (NA)

Text File Formats: Data content associated with text files (i.e., .txt, .doc, .pdf) was found only for .txt files when sent via email or the BlackBerry desktop manager protocol. No deleted data was found. (Below)

Graphics File Formats: Graphics file (e.g., .bmp, .jpg, .gif, .png, .tif) data content was not found or displayed when sent via email. The BlackBerry desktop manager protocol does not allow for transferring image files to the device. The filename and subject line of the email were found and reported. (Miss)

Compressed File Archive Formats: Compressed data file (i.e., `.zip`, `.rar`, `.exe`, `.tgz`) content was not found. (Miss)

Misnamed Files: Misnamed file (e.g., `.txt` file renamed with a `.dll` extension) data content was not found. (Miss)

Peripheral Memory Cards: N.A. – The BlackBerry 7750 does not allow for removable media. (NA)

Acquisition Consistency: Two consecutive acquisition produce different overall hashes of the memory. However, the individual database files acquired were consistent. (Meet)

Cleared Devices: Two approaches exist for performing a hard reset on a BlackBerry device: 1) Not supplying the correct pass-phrase within 10 attempts 2) Accessing the BlackBerry Desktop Manager and clearing all databases via the Backup/Restore advanced menu. The first approach forces the user to re-download the OS before the device is functional. No data was found. (Meet)

The following data was found after clearing the device through the desktop manager: PIM data (i.e., Address Book, Calendar), dialed/received phone calls and Internet Messaging data (i.e., subject/filename). (Above)

Power Loss: The BlackBerry 7750 was repopulated with the above scenarios, then completely drained of all battery power and reacquired. All data was found as reported above. (Above)

BlackBerry 7780

The following scenarios were executed on an AT&T GSM BlackBerry 7780 Java-based wireless handheld device running v3.7.1.59 (Platform 1.6.1.48). PDA Seizure version 3.0.1.35 was used for acquisition.

Connectivity and Retrieval: The device contents were successfully acquired with or without the SIM present. If authentication mechanisms are applied, the proper pass-phrase must be provided within 10 attempts. The correct pass-phrase must be provided for both Memory and Database acquisitions. Basic subscriber and service provider information (e.g., IMEI, ICCID, MSISDN) were not found. The reported memory size is inconsistent with the total size found on the device via the Options->Status screen. (Meet)

PIM Applications: All PIM data was found and reported (i.e., Address Book, Calendar, Tasks, Memos) in the corresponding databases (i.e., Address Book, Calendar, Tasks, Memos database). All deleted PIM data was found and reported in the memory window. (Meet)

Dialed/Received Phone Calls: All dialed/received phone calls were found and reported in the Phone Call Log, Phone Hotlist database, and the memory window when conducting a search. Deleted calls were found in the memory window. (Meet)

SMS/MMS Messaging: All active incoming and outgoing SMS messages were found and reported in the Messages folder and the memory window when conducting a search on message content. All deleted SMS messages were found and reported in the memory window. MMS messages are not supported. (Meet)

Internet Messaging: All data content associated with sent and received email messages was found and reported in the Messages database and the memory window. Deleted messages were found and reported in the memory window. The BlackBerry device does not support Instant Messaging and, therefore omitted from this scenario. (Meet)

Web Applications: Visited URLs and search engine queries were found and reported. Textual Web content pertaining to URLs was not found. No graphical images of visited sites were found and displayed. (Below)

Text File Formats: Data content associated with text files (i.e., .txt, .doc, .pdf) was found only for .txt files when sent via email or the BlackBerry desktop manager protocol. The filename and subject line of the email were found and reported for all text-based files. No deleted data was found. (Below)

Graphics File Formats: Graphics file (e.g., .bmp, .jpg, .gif, .png, .tif) data content was not found or displayed when sent via email. The BlackBerry desktop manager protocol does not allow for transferring image files to the device. The filename and subject line of the email were found and reported. (Miss)

Compressed File Archive Formats: Compressed data file (i.e., .zip, .rar, .exe, .tgz) content was not found. (Miss)

Misnamed Files: Misnamed file (e.g., .txt file renamed with a .dll extension) data content was not found. (Miss)

Peripheral Memory Cards: N.A. – The BlackBerry 7780 does not allow for removable media. (NA)

Acquisition Consistency: Two consecutive acquisition produce different overall hashes of the memory. The individual database files acquired were consistent. (Meet)

Cleared Devices: Two approaches exist for performing a hard reset on a BlackBerry device: 1) Not supplying the correct pass-phrase within 10 attempts 2) Accessing the BlackBerry Desktop Manager and clearing all databases via the Backup/Restore advanced menu. The first approach forces the user to re-download the OS before the device is functional. No data was found. (Meet)

The following data was found after clearing the device through the desktop manager: PIM data (i.e., Address Book, Calendar), dialed/received phone calls and Internet Messaging data (i.e., subject/filename). (Above)

Power Loss: The BlackBerry 7780 was repopulated with the above scenarios, then completely drained of all battery power and reacquired. All data was found as reported above. (Above)

Kyocera 7135

The following scenarios were conducted on a Verizon Kyocera 7135 running Palm OS version 4.1. PDA Seizure version 3.0.3.89 was used for acquisition.

Connectivity and Retrieval: The device contents were successfully acquired. If authentication mechanisms are applied, the proper pass-phrase has to be provided in order to begin acquisition. Basic subscriber and service provider information (e.g., ESN, MSID, FCCID) were not found. Network carrier and the phone number was found in the NetworkDB file and reported. Since internal memory could not be acquired, memory size was not reported and could not be verified. (Meet)

PIM Applications: All PIM data was found (i.e., Address Book, Calendar, Tasks, Memos) in the corresponding database/folder (i.e., AddressDB, DatebookDB, ToDoDB, MemoDB files) and reported. The following deleted PIM data was found and reported in association to the Address Book entries: Company Name, found in the AddressCompaniesDB, and Titles, found in the AddressTitlesDB. (Meet)

Dialed/Received Phone Calls: All dialed/received phone calls were found and reported in the kwc_CallHistoryDB file. Deleted phone calls were found in the memory dump. (Meet)

SMS/MMS Messaging: All active incoming and outgoing SMS messages were found in the kwc_messages file and reported. Deleted SMS messages were not found or reported. MMS Messaging is not supported. (Below)

Internet Messaging: All data content associated with sent and received email messages was found in the MailDB, pdQmailMsgs and the MMPROIII Message files and reported. Deleted messages were not found. (Below)

Web Applications: Visited URLs were found and reported, while search queries performed were not found. Textual or graphical Web content pertaining to URLs was not found. (Below)

Text File Formats: Data content associated with text files (i.e., .txt, .doc, .pdf) was found and reported for .txt files. However, .pdf and .doc file content sent via email and Hotsync was not found in a readable format. Necessary software that allows each of the above file types to be read was installed on the device. Deleted text file data was found and reported for .txt files. (Meet)

Graphics File Formats: Graphic file (i.e., .bmp, .jpg, .gif, .png, .tif) data was not found when sent via email or Hotsync. (Miss)

111

Compressed File Archive Formats: Compressed data file (i.e., `.zip`, `.rar`, `.exe`, `.tgz`) content was not found. (Miss)

Misnamed Files: Misnamed file (e.g., `.txt` file renamed with a `.dll` extension) data content was found and reported for text-based files, when sent via email. Unknown file types are not accepted by the Hotsync protocol. (Meet)

Peripheral Memory Cards: No data residing on a 128 MB MMC populated with various files (i.e., text, graphics, audio, compressed archive files, misnamed files) was found. (Miss)

Acquisition Consistency: Two consecutive acquisition produce different hashes on the following database files: NetworkDB and Saved Preferences. (Below)

Cleared Devices: A Hard Reset was performed by, holding the reset button while pressing the illumination key. No data was found. (Meet)

Power Loss: The Kyocera 7135 was repopulated with the above scenarios, then completely drained of all battery power and reacquired. No data was found. (Meet)

Motorola MPx220

The following scenarios were conducted on a Cingular GSM Motorola MPx220 running Microsoft Windows Mobile 2004 for Pocket PC Phone Edition. PDA Seizure version 3.0.3.89 was used for acquisition.

Connectivity and Retrieval: The data contents of the device were successfully acquired with or without the SIM present. If internal memory authentication mechanisms are applied, the proper pass-phrase must be provided in order for the ActiveSync connection to take place. Basic subscriber and service provider information (e.g., IMEI, ICCID, MSISDN) were not found. Network carrier and the phone number was found in the \Windows directory and reported. The reported memory size is consistent with the total memory size of the device. (Meet)

PIM Applications: PIM data was not found (i.e., Address Book, Calendar, Tasks, Memos). (Miss)

Dialed/Received Phone Calls: Dialed/received phone calls were not found. (Miss)

SMS/MMS Messaging: SMS messages were not found. Textual content of all active incoming MMS messages were found and reported. MMS attachments were not found. Deleted MMS messages were not found. (Below)

Internet Messaging: All data content associated with sent and received email messages was found and reported. Deleted messages were not found. (Below)

Web Applications: Visited URLs and search engine queries were found and reported. Textual Web content pertaining to visited URLs was not found. No graphical images of visited sites were found. (Below)

Text File Formats: Data content associated with text files (i.e., .txt, .doc, .pdf) was found and reported. Deleted text files were not found. (Below)

Graphics File Formats: Graphics file (i.e., .bmp, .jpg, .gif, .png, .tif) data content was found and reported. .png files were found but not displayed in the graphics library. Deleted graphics files were not found. (Below)

Compressed File Archive Formats: Compressed data file (i.e., .zip, .rar, .exe, .tgz) content was found and reported. (Meet)

Misnamed Files: Misnamed file (e.g., .txt file renamed with a .dll extension) data content was found and reported. (Meet)

Peripheral Memory Cards: Data residing on a 256 MB Mini SD Card populated with various files (i.e., text, graphics, audio, compressed archive files, misnamed files) was found and reported. Deleted files were not found. (Below)

Acquisition Consistency: Two consecutive acquisition produce different overall hashes of the memory. The individual files acquired were consistent. (Meet)

Cleared Devices: A Hard Reset was performed by holding down the action button and pressing the power button. No data was found. (Meet)

Power Loss: The Motorola MPx220 was repopulated with the above scenarios, then completely drained of all battery power and reacquired. The following data was found: MMS messages (text content) and Internet Messaging data (i.e., sent/received email). Individual files stored in the /storage directory were found and reported. (Above)

Samsung i700

The following scenarios were conducted on a Verizon CDMA Samsung i700 running Microsoft Windows Mobile 2004 for Pocket PC Phone Edition. PDA Seizure version 3.0.3.89 was used for acquisition.

Connectivity and Retrieval: The device contents were successfully acquired. If authentication mechanisms are applied, the proper pass-phrase must be provided in order for the ActiveSync connection to take place. Basic subscriber and service provider information (e.g., ESN, MSID, FCCID) were not found. Network carrier and the phone number was found in the \Windows directory and reported. The reported memory size is consistent with the total memory size of the device. (Meet)

PIM Applications: All PIM data was found (i.e., Address Book, Calendar, Tasks, Memos) in the corresponding database/folder (i.e., Contacts, Appointments, Tasks database, Memos) and reported. Deleted PIM data was found and reported. (Meet)

Dialed/Received Phone Calls: All dialed/received phone calls were found in the clog database and the memory image and reported. Phone numbers were found by issuing the following search pattern: (.a.a.a)..p.p.p.-.s.s.s.s. Deleted calls were found in the memory image. (Meet)

SMS/MMS Messaging: All active incoming and outgoing SMS messages were found in the \Windows\Messaging folder and the fldr100171c file and reported. Deleted SMS messages were not found. MMS messaging is not supported. (Below)

Internet Messaging: All data content associated with sent and received email messages was found in the \Windows\Messaging file and reported. Deleted message file content (e.g., subject, body text) was not found. (Below)

Web Applications: Visited URLs and search engine queries were found and reported. Textual Web content pertaining to URLs was found and reported. No graphical images of visited sites were found or displayed. (Below)

Text File Formats: Data content associated with text files (i.e., .txt, .doc, .pdf) was found and reported, when transferred via email and ActiveSync. Deleted text file data was not found. (Below)

Graphics File Formats: Graphic file (i.e., .bmp, .jpg, .gif, .png, .tif) filenames and data content were found and displayed when sent via email and ActiveSync. Deleted graphic file data was not found. (Below)

Compressed File Archive Formats: Compressed data (i.e., .zip, .rar, .exe, .tgz) filenames and content were found and reported when sent via email and ActiveSync. (Meet)

Misnamed Files: Misnamed file (e.g., .txt file renamed with a .dll extension) data content was found when sent via email and ActiveSync. (Meet)

Peripheral Memory Cards: Data residing on a 128 MB MMC populated with various files (i.e., text, graphics, audio, compressed archive files, misnamed files) was found and reported. Deleted files were not found. (Below)

Acquisition Consistency: Two consecutive acquisition produce different hashes on the following database files: \Categories Database, \ConfigMetabase, SchedSync.dat, DB_notify_events, DB_notify_queue, pmailFolders, pmailMsgClasses, pmailNamedProps, pmailServices, and Speed.db. (Below)

Cleared Devices: A Hard Reset was performed by holding down the power button while pressing the reset button with the stylus, then releasing the power button. No data was found. (Meet)

Power Loss: The Samsung i700 Pocket PC was repopulated with the above scenarios and completely drained of all battery power and reacquired. No data was found. (Meet)

PalmOne Treo 600

The following scenarios were conducted on the PalmOne Treo 600 running Palm OS version 5.2.1. PDA Seizure version 3.0.3.89 was used for acquisition.

Connectivity and Retrieval: The device contents were successfully acquired with or without the SIM present. When authentication mechanisms were applied, the proper pass-phrase had to be provided in order to begin acquisition. Basic subscriber and service provider information (e.g., IMEI, ICCID, MSISDN) were not found. The memory size was not reported and could not be verified. (Meet)

PIM Applications: All active PIM data was found (i.e., Address Book, Calendar, Tasks, Memos) in the corresponding database/folder (i.e., AddressDB, DatebookDB, ToDoDB, MemoDB files) and reported. Deleted PIM data was found in the memory file. (Meet)

Dialed/Received Phone Calls: All dialed/received phone calls were found in the PhoneCallDB file and reported. Deleted phone calls were found in the memory file. (Meet)

SMS/MMS Messaging: All active SMS messages were found and reported in the Msg Database.pdb file. MMS messages and corresponding filenames were found and reported. However, the associated multi-media files were not viewable. Deleted SMS/MMS messages were found in the memory file. (Meet)

Internet Messaging: All data content associated with sent and received email messages was found in the Email_libr_HsMp_BDC79AAB file and reported. Deleted message file content (e.g., subject, body text) was not found, but the email address was found in the EmailAddressDB.pdb file. (Below)

Web Applications: Visited URLs and search engine queries were found and reported. Textual Web content pertaining to URLs was found and reported. No graphical images of visited sites were found or displayed. (Below)

Text File Formats: Data content associated with text files (i.e., .txt, .doc, .pdf) was found and reported for .doc and .txt files. However, .pdf file content data sent via email and Hotsync was not found in a readable format. Necessary software that allows each of the above file types to be read was installed on the device. Deleted text file data was found and reported for .doc and .txt files. (Meet)

Graphics File Formats: Graphic file (i.e., .bmp, .jpg, .gif, .png, .tif) data was not found when sent via email or Hotsync. (Miss)

115

Compressed File Archive Formats: Compressed data file (i.e., .zip, .rar, .exe, .tgz) content was not found, but the filename embedded in the compressed file was found and reported when sent via email. (Below)

Misnamed Files: Misnamed file (e.g., .txt file renamed with a .dll extension) data content was found and reported for text-based files, but only when sent via email. Unknown file types are not accepted by the Hotsync protocol. (Meet)

Peripheral Memory Cards: No data residing on a 128 MB MMC populated with various files (i.e., text, graphics, audio, compressed archive files, misnamed files) was found or reported. (Miss)

Acquisition Consistency: Two consecutive acquisition produce different hashes on numerous database files. (Below)

Cleared Devices: A Hard Reset was performed by holding down the reset button and pressing the power key. No data was found. (Meet)

Power Loss: The Treo 600 was repopulated with the above scenarios, then completely drained of all battery power and reacquired. No data was found. (Meet)

Appendix B: Pilot-Link Results

The scenarios were performed on a Forensic Recovery of Evidence Device (FRED) running RedHat Linux. Pilot-Link version 0.11.8 was used to acquire data from Palm OS devices with cell phone capabilities. Two Pilot-Link commands were used to acquire data: `pi-getram` (acquires internal memory) and `pilot-xfer -b` (acquires individual databases).

Kyocera 7135

The following scenarios were conducted on a Verizon Kyocera 7135 running Palm OS version 4.1.

Connectivity and Retrieval: The device contents were successfully acquired. If authentication mechanisms are applied, the proper pass-phrase must be provided in order to begin acquisition. Basic subscriber and service provider information (e.g., ESN, MSID, FCCID) were not found. The reported memory size is consistent with the total size found on the device. (Meet)

PIM Applications: All PIM data was found (i.e., Address Book, Calendar, Tasks, Memos) in the corresponding database/folder (i.e., AddressDB, DatebookDB, ToDoDB, MemoDB files) and reported. All deleted PIM data was found and reported in the memory image file – pilot-4.1.2.ram: Address Book, Calendar, Tasks and Memo entries. (Meet)

Dialed/Received Phone Calls: All dialed/received phone calls were found and reported in the kwc_CallHistoryDB file. All deleted phone calls were found in the pilot-4.1.2 memory image file. (Meet)

SMS/MMS Messaging: All active incoming and outgoing SMS messages were found in the kwc_messages file and reported. Deleted SMS messages were found and reported in the pilot-4.1.2 memory image file. MMS Messaging is not supported. (Meet)

Internet Messaging: All data content associated with sent and received email messages was found in the pdQmailMsgs and the pdQmailTOC.pdb files and reported. Deleted messages were found and reported in the pilot-4.1.2 memory image file. (Meet)

Web Applications: Visited URLs and search queries performed were found and reported. Textual Web content pertaining to URLs was found and reported. No graphical images of visited sites were found and displayed. (Below)

Text File Formats: Data content associated with text files (i.e., .txt, .doc, .pdf) was found and reported for .txt and .doc files. However, .pdf file content sent via email and Hotsync was not found in a readable format. Necessary software that allows each of the above file types to be read was installed on the device. Deleted text file data was found and reported for .txt and .doc files. (Meet)

Graphics Files Format: Graphic files (e.g., .bmp, .jpg, .gif, .png, .tif) filenames were found but not displayed when sent via email and Hotsync. Necessary graphics software that allows

117

each of the above file types to be viewed was installed on the device. Deleted graphic filenames were found and reported. (Below)

Compressed File Archive Formats: Compressed data file (i.e., .zip, .rar, .exe, .tgz) content was not found. (Miss)

Misnamed Files: Misnamed file (e.g., .txt file renamed with a .dll extension) data content was found and reported for text-based files, but only when sent via email. Unknown file types are not accepted by the Hotsync protocol. (Meet)

Peripheral Memory Cards: No data residing on a 128 MB MMC populated with various files (i.e., text, graphics, audio, compressed archive files, misnamed files) was found. (Miss)

Acquisition Consistency: The consistencies of consecutive acquisitions were determined by SHA and MD5 hashing algorithms. The following database files produced inconsistent checksums: Net Prefs.prc, NetworkDB.pdb, Saved Prefrerences.prc. (Below)

Cleared Devices: A Hard Reset was performed by removing the battery cover and holding the reset button while pressing the illumination key. No data was found. (Meet)

Power Loss: The Kyocera 7135 was repopulated with the above scenarios, then completely drained of all battery power and reacquired. No data was found. (Meet)

PalmOne Treo 600

The following scenarios were conducted on the PalmOne Treo 600 running Palm OS version 5.2.1.

Connectivity and Retrieval: The device contents were partially acquired. A physical memory image of the device could not be acquired. Individual database files were successfully acquired with or without the SIM present. When authentication mechanisms were applied, the proper pass-phrase had to be provided in order to begin acquisition. Basic subscriber and service provider information (e.g., IMEI, ICCID, MSISDN) were not found. (Below)

PIM Applications: All active PIM data was found (i.e., Address Book, Calendar, Tasks, Memos) in the corresponding database/folder (i.e., AddressDB, DatebookDB, ToDoDB, MemoDB files) and reported. The following deleted PIM data related to Address Book entries were found and reported: Company Name (found in the AddressCompaniesDB) and Titles (found in the AddressTitlesDB). (Meet)

Dialed/Received Phone Calls: All dialed/received phone calls were found in the PhoneCallDB file and reported. Deleted phone calls were not found. (Below)

SMS/MMS Messaging: All active incoming and outgoing SMS messages were found in the Msg Database.pdb file and reported. Deleted SMS messages were not found. Incoming and outgoing MMS messages and corresponding filenames were found and reported. However, the associated multi-media files were not viewable. Deleted MMS messages were not found or reported. (Below)

Internet Messaging: All data content associated with sent and received email messages was found in the Email_libr_HsMp pdb file and reported. Deleted message file content (e.g., subject, body text) was not found, but the email address was found in the EmailAddressDB.pdb file. (Below)

Web Applications: Visited URLs and search engine queries were found and reported. Textual Web content pertaining to URLs was found and reported. No graphical images of visited sites were found or displayed. (Below)

Text File Formats: Data content associated with text files (i.e., .txt, .doc, .pdf) was found and reported for .doc and .txt files. However, .pdf file content data sent via email and Hotsync was not found in a readable format. Necessary software that allows each of the above file types to be read was installed on the device. Deleted text file data was found and reported for .doc and .txt files. (Meet)

Graphics Files Format: Graphic files (e.g., .bmp, .jpg, .gif, .png, .tif) filenames were found but not displayed when sent via email and Hotsync. Necessary graphics software that allows each of the above file types to be viewed was installed on the device. Deleted graphic filenames were found and reported. (Below)

Compressed File Archive Formats: Compressed data file (i.e., .zip, .rar, .exe, .tgz) content was not found. (Miss)

Misnamed Files: Misnamed file (e.g., .txt file renamed with a .dll extension) data content was found and reported for text-based files when sent via email. Unknown file types are not accepted by the Hotsync protocol. (Meet)

Peripheral Memory Cards: No data residing on a 128 MB MMC populated with various files (i.e., text, graphics, audio, compressed archive files, misnamed files) was found or reported. (Miss)

Acquisition Consistency: The consistencies of consecutive acquisitions were determined by SHA and MD5 hashing algorithms. The following database files produced inconsistent checksums: Email_HsDm_appl_a68k.prc, Email_libr_HsMp_BDC79AAB.pdb, Graffiti Shortcuts.prc, HSTraceDatabase.pdb, Launcher_lnch_appl_a68k.prc, PhoneCallDB.pdb, Phone_HsPh_appl_a68k.prc. (Below)

Cleared Devices: A Hard Reset was performed by holding down the reset button and pressing the power key. No data was found. (Meet)

Power Loss: The Treo 600 was repopulated with the above scenarios, then completely drained of all battery power and reacquired. No data was found. (Meet)

119

Appendix C: Cell Seizure Results

The scenarios were performed on a Forensic Recovery of Evidence Device (FRED) running Windows XP SP2. Cell Seizure version 2.0.0.336660 was used to acquire data from the following cell phones: Ericsson T68i, Motorola C333, Motorola V66, Nokia 3390, and the Nokia 6610i.

Ericsson T68i

The following scenarios were conducted on a Sony Ericsson T68i. Cell Seizure version 2.0.0.33660 was used for acquisition.

Connectivity and Retrieval: Data-Pilot's Susteen data cable for Sony Ericsson phones was used in order to establish connectivity with Cell Seizure. Proper authentication had to be provided to the password-protected device and the SIM card had to be inserted before contents were successfully acquired. Basic subscriber and service provider information was found and reported (i.e., IMEI). Memory size is not reported. (Meet)

PIM Applications: All PIM data was found and reported (i.e., Address Book, Calendar, Tasks). Deleted PIM data was not found. (Below)

Dialed/Received Phone Calls: All dialed/received phone calls were found and reported in the Phone calls folder. Deleted phone calls were not found. (Below)

SMS/MMS Messaging: All active incoming and outgoing SMS messages were found in the SMS folder and reported. Deleted SMS messages were not found. MMS Messages and attachments (i.e., graphics, sound bytes) were not found or reported. (Below)

Internet Messaging: Data content associated with sent and received email messages were not found. (Miss)

Web Applications: Visited URLs, search queries performed, textual Web content or graphical images of visited sites were not found. (Miss)

Text File Formats: N.A. – The Sony Ericsson T68i does not support text files (e.g., .txt, .doc, .pdf). (NA)

Graphics Files Format: Supported graphic files (i.e., .jpg, .gif) present on the device were not found. (Miss)

Compressed File Archive Formats: N.A. – The Sony Ericsson T68i does not support compressed archive files (e.g., .zip, .rar, .exe, .tgz). (NA)

Misnamed Files: N.A. – The Sony Ericsson T68i does not support misnamed files (e.g., .txt file renamed with a .dll extension). (NA)

Peripheral Memory Cards: N.A. – The Sony Ericsson T68i does not allow for removable media. (NA)

Acquisition Consistency: All hashes of individual folders were consistent. (Meet)

Cleared Devices: A Hard Reset was performed by selecting, Master Reset in the settings menu. SMS messages and quicknotes were found in the memory dump file. (Above)

Power Loss: The Sony Ericsson T68i was repopulated with the above scenarios, then completely drained of all battery power and reacquired. All data was found as reported above. (Above)

Motorola C333

The following scenarios were conducted on an unlocked GSM Motorola C333. Cell Seizure version 2.0.0.33660 was used for acquisition.

Connectivity and Retrieval: The password-protected device contents were successfully acquired with or without the SIM present without having to provide proper authentication. Basic subscriber and service provider information (i.e., IMEI) was found. Memory size is not reported. (Meet)

PIM Applications: All PIM data was found (i.e., Address Book, Calendar) in the corresponding database/folder (i.e., Phonebook, Datebook folders) and reported. Deleted PIM data was found and reported in the Filesystem folder. (Meet)

Dialed/Received Phone Calls: All dialed/received phone calls were found and reported in the Call History folder. Deleted phone calls were not found. (Below)

SMS/MMS Messaging: All active incoming and outgoing SMS messages were found in the SMS Message folder and reported. Deleted outgoing SMS messages were found in the SMS and Quick Notes dump folder. Deleted incoming SMS messages were not found. Incoming/outgoing MMS Messages were not found. (Meet)

Internet Messaging: N.A. – The Motorola C333 does not support email. (NA)

Web Applications: N.A. – Internet connectivity was unable to be established. (NA)

Text File Formats: N.A. – The Motorola C333 does not support text files (e.g., .txt, .doc, .pdf). (NA)

Graphics Files Format: N.A. – The Motorola C333 does not support graphic files (e.g., .bmp, .jpg, .gif, .png, .tif). (NA)

Compressed File Archive Formats: N.A. – The Motorola C333 does not support compressed archive files (e.g., .zip, .rar, .exe, .tgz). (NA)

121

Misnamed Files: N.A. – The Motorola C333 does not support misnamed files (e.g., .txt file renamed with a .dll extension). (NA)

Peripheral Memory Cards: N.A. – The Motorola C333 does not allow for removable media. (NA)

Acquisition Consistency: All hashes of individual folders were consistent. (Meet)

Cleared Devices: A Hard Reset was performed by selecting the Master Clear and Master Reset in the settings menu. SMS messages and Memos were found and reported in the SMS and Quick notes dump folder. (Above)

Power Loss: The Motorola C333 was repopulated with the above scenarios, then completely drained of all battery power and reacquired. All data was found as reported above. (Above)

Motorola V66

The following scenarios were conducted on an unlocked GSM Motorola V.series 66. Cell Seizure version 2.0.0.33660 was used for acquisition.

Connectivity and Retrieval: The password-protected device contents were successfully acquired with or without the SIM present without having to provide proper authentication. Basic subscriber and service provider information (i.e., IMEI, ICCID, IMSI) was found. Memory size is not reported. (Meet)

PIM Applications: All PIM data was found (i.e., Address Book, Calendar) in the corresponding database/folder (i.e., Phonebook, Datebook folders) and reported. Deleted PIM data was not found. (Below)

Dialed/Received Phone Calls: All dialed/received phone calls were found and reported in the Call History folder. Deleted phone calls were not found. (Below)

SMS/MMS Messaging: All active incoming and outgoing SMS messages were found in the SMS Message folder and reported. Deleted outgoing SMS messages were found in the SMS and Quick Notes dump folder. MMS Messaging is not supported. (Meet)

Internet Messaging: N.A. – The Motorola V66 does not support email. (NA)

Web Applications: N.A. – Internet connectivity was unable to be established. (NA)

Text File Formats: N.A. – The Motorola V66 does not support text files (e.g., .txt, .doc, .pdf). (NA)

Graphics Files Format: N.A. – The Motorola V66 does not support graphic files (e.g., .bmp, .jpg, .gif, .png, .tif). (NA)

Compressed File Archive Formats: N.A. – The Motorola V66 does not support compressed archive files (e.g., .zip, .rar, .exe, .tgz). (NA)

Misnamed Files: N.A. – The Motorola V66 does not support misnamed files (e.g., .txt file renamed with a .dll extension). (NA)

Peripheral Memory Cards: N.A. – The Motorola V66 does not allow for removable media. (NA)

Acquisition Consistency: The following database file produced inconsistent checksums: SMS and Quick notes dump. (Below)

Cleared Devices: A Hard Reset was performed by selecting, Master Clear/Reset in the settings menu. No data was found. (Meet)

Power Loss: The Motorola V66 was repopulated with the above scenarios, then completely drained of all battery power and reacquired. All data was found as reported above. (Above)

Nokia 3390

The following scenarios were conducted on a GSM Nokia 3390. Cell Seizure version 2.0.0.33660 was used for acquisition.

Connectivity and Retrieval: The password-protected device contents were successfully acquired with or without the SIM present providing proper authentication. Basic subscriber and service provider information was found (i.e., IMEI). Memory size is not reported. (Meet)

PIM Applications: All PIM data was found and reported (i.e., Phonebook, Calendar). Deleted PIM data was not found. (Below)

Dialed/Received Phone Calls: All dialed/received phone calls were found and reported in the Phone calls folder. Deleted phone calls were not found. (Below)

SMS/MMS Messaging: All active incoming SMS messages were found in the SMS folder and reported. Outgoing SMS messages were not found. Deleted SMS messages were not found. MMS Messages are not supported. (Below)

Internet Messaging: N.A. – The Nokia 3390 does not support email. (NA)

Web Applications: The Nokia 3390 does not support browsing the Web but allows for Instant Messaging. No data was found. (Miss)

Text File Formats: N.A. – The Nokia 3390 does not support text files (e.g., .txt, .doc, .pdf). (NA)

Graphics Files Format: N.A. – The Nokia 3390 does not support graphic files (e.g., .bmp, .jpg, .gif, .png, .tif). (NA)

Compressed File Archive Formats: N.A. – The Nokia 3390 does not support compressed archive files (e.g., `.zip`, `.rar`, `.exe`, `.tgz`). (NA)

Misnamed Files: N.A. – The Nokia 3390 does not support misnamed files (e.g., `.txt` file renamed with a `.dll` extension). (NA)

Peripheral Memory Cards: N.A. – The Nokia 3390 does not allow for removable media. (NA)

Acquisition Consistency: All hashes of individual folders were consistent. (Meet)

Cleared Devices: N.A. – A Hard Reset function is not provided by the phone. (NA)

Power Loss: The Nokia 3390 was repopulated with the above scenarios, then completely drained of all battery power and reacquired. All data was found as reported above. (Above)

Nokia 6610i

The following scenarios were conducted on an unlocked GSM Nokia 6610i. Cell Seizure version 2.0.0.33660 was used for acquisition.

Connectivity and Retrieval: Proper authentication had to be provided to the password-protected device when the SIM was present, although acquisition was successful without the SIM. Basic subscriber and service provider information was found and reported (i.e., IMEI). Memory size is not reported. (Meet)

PIM Applications: All PIM data was found and reported (i.e., Phonebook, Calendar). Deleted PIM data was not found. (Below)

Dialed/Received Phone Calls: All dialed/received phone calls were found and reported in the Call Logs folder. Deleted phone calls were not found. (Below)

SMS/MMS Messaging: All active incoming and outgoing SMS messages were found in the SMS History folder and reported. Active MMS messages and associated files (e.g., graphic, audio files) were found. Deleted SMS/MMS messages were not found. (Below)

Internet Messaging: N.A. – The Nokia 6610i does not support email. (NA)

Web Applications: Visited URLs were found and reported, while search queries performed were not found. Textual Web content pertaining to visited URLs was not found. No graphical images of visited sites were found and displayed. (Below)

Text File Formats: Data content associated with text files (i.e., `.txt`, `.doc`, `.pdf`) was found and reported in the Filesystem folder and could be viewed after saved to the forensic workstation. Deleted text files were not found. (Below)

Graphics Files Format: Graphic files (e.g., .bmp, .jpg, .gif, .png, .tif) present on the device were found and reported. Deleted graphic files were not found. (Below)

Compressed File Archive Formats: Compressed data file (i.e., .zip, .rar, .exe, .tgz) content was found and reported in the Files folder and could be viewed after saved to the forensic workstation. (Meet)

Misnamed Files: Misnamed files (e.g., .txt file renamed with a .dll extension) were found and reported in the Files folder and could be viewed with the proper application after saved to the forensic workstation. (Meet)

Peripheral Memory Cards: N.A. – The Nokia 6610i does not allow for removable media. (NA)

Acquisition Consistency: All hashes of individual folders were consistent. (Meet)

Cleared Devices: N.A. – A Hard Reset function is not provided by the phone. (NA)

Power Loss: The Nokia 6610i was repopulated with the above scenarios, then completely drained of all battery power and reacquired. All data was found as reported above. (Above)

Appendix D: GSM .XRY Results

The scenarios were performed on a Forensic Recovery of Evidence Device (FRED) running Windows XP SP2. GSM .XRY version 2.5 was used to acquire data from the following cell phones: Ericsson T68i, Motorola V66, Motorola V300, Nokia 6610i, Nokia 6200, and the Nokia 7610.

Ericsson T68i

The following scenarios were conducted on a Sony Ericsson T68i. GSM .XRY version 2.5 was used for acquisition.

Connectivity and Retrieval: Proper authentication had to be provided to the password-protected device and the SIM card had to be inserted before contents were successfully acquired. Basic subscriber and service provider information was found and reported (i.e., IMEI, IMSI). Memory size is not reported. (Meet)

PIM Applications: All PIM data was found (i.e., Address Book, Calendar, Tasks) in the corresponding database/folder (i.e., Contacts, Calendar folders, Tasks) and reported. Deleted PIM data was not found. (Below)

Dialed/Received Phone Calls: All dialed/received phone calls were found and reported in the Calls folder. Deleted phone calls were not found. (Below)

SMS/MMS Messaging: All active incoming and outgoing SMS messages were found in the SMS folder and reported. Deleted SMS messages were not found. MMS Messages and attachments (i.e., graphics, sound bytes) were not found or reported. (Below)

Internet Messaging: Data content associated with sent and received email messages, was not found. (Miss)

Web Applications: Visited URLs, search queries performed, textual Web content or graphical images of visited sites were not found. (Miss)

Text File Formats: N.A. – The Sony Ericsson T68i does not support text files (e.g., .txt, .doc, .pdf). (NA)

Graphics Files Format: Supported graphic files (i.e., .jpg, .gif) present on the device were not found. (Miss)

Compressed File Archive Formats: N.A. – The Sony Ericsson T68i does not support compressed archive files (e.g., .zip, .rar, .exe, .tgz). (NA)

Misnamed Files: N.A. – The Sony Ericsson T68i does not support misnamed files (e.g., .txt file renamed with a .dll extension). (NA)

Peripheral Memory Cards: N.A. – The Sony Ericsson T68i does not allow for removable media. (NA)

Acquisition Consistency: N.A. – This signature is not a hash of the device data, but the identity of the examiner. (NA)

Cleared Devices: A Hard Reset was performed by selecting, Reset Settings and Reset All, in the settings menu. No data was found. (Meet)

Power Loss: The Sony Ericsson T68i was repopulated with the above scenarios, then completely drained of all battery power and reacquired. All data was found as reported above. (Above)

Motorola V66

The following scenarios were conducted on an unlocked GSM Motorola V.series 66. GSM .XRY version 2.5 was used for acquisition.

Connectivity and Retrieval: The password-protected device contents were successfully acquired with or without the SIM present without having to provide proper authentication. Basic subscriber and service provider information was found and reported (i.e., IMEI, IMSI). Memory size is not reported. (Meet)

PIM Applications: All PIM data was found (i.e., Address Book, Calendar) in the corresponding database/folder (i.e., Contacts, Calendar folders) and reported. Deleted PIM data was not found. (Below)

Dialed/Received Phone Calls: All dialed/received phone calls were found and reported in the Calls folder. Deleted phone calls were not found. (Below)

SMS/MMS Messaging: All active incoming and outgoing SMS messages were found in the SMS folder and reported. Deleted SMS messages were not found. MMS Messaging is not supported. (Below)

Internet Messaging: N.A. – The Motorola V66 does not support email. (NA)

Web Applications: N.A. – Internet connectivity was unable to be established. (NA)

Text File Formats: N.A. – The Motorola V66 does not support text files (e.g., .txt, .doc, .pdf). (NA)

Graphics Files Format: N.A. – The Motorola V66 does not support graphic files (e.g., .bmp, .jpg, .gif, .png, .tif). (NA)

Compressed File Archive Formats: N.A. – The Motorola V66 does not support compressed archive files (e.g., .zip, .rar, .exe, .tgz). (NA)

Misnamed Files: N.A. – The Motorola V66 does not support misnamed files (e.g., .txt file renamed with a .dll extension). (NA)

Peripheral Memory Cards: N.A. – The Motorola V66 does not allow for removable media. (NA)

Acquisition Consistency: N.A. – This signature is not a hash of the device data, but the identity of the examiner. (NA)

Cleared Devices: A Hard Reset was performed by selecting, Master Reset, in the settings menu. No data was found. (Meet)

Power Loss: The Motorola V66 was repopulated with the above scenarios, then completely drained of all battery power and reacquired. All data was found as reported above. (Above)

Motorola V300

The following scenarios were conducted on a Pay-As-You-Go GSM Motorola V300. GSM .XRY version 2.5 was used for acquisition.

Connectivity and Retrieval: Proper authentication had to be provided to the password-protected device and the SIM card had to be inserted before contents were successfully acquired. Basic subscriber and service provider information was found and reported (i.e., IMEI, IMSI). Memory size is not reported. (Meet)

PIM Applications: All PIM data was found (i.e., Address Book, Calendar) in the corresponding database/folder (i.e., Contacts, Calendar folders) and reported. Deleted PIM data was not found. (Below)

Dialed/Received Phone Calls: All dialed/received phone calls were found and reported in the Calls folder. Deleted phone calls were not found. (Below)

SMS/MMS Messaging: All active incoming and outgoing SMS messages were found in the SMS folder and reported. Deleted SMS messages were not found. MMS Messages and attachments (i.e., graphics, sound bytes) were not found or reported. (Below)

Internet Messaging: Incoming/outgoing emails were found and reported. Chat logs and Deleted messages were not found. (Below)

Web Applications: Visited URLs, search queries performed, textual Web content or graphical images of visited sites were not found. (Miss)

Text File Formats: Data content associated with text files (i.e., .txt, .doc, .pdf) was not found when sent via email. The filename and subject line of the email were found and reported. (Miss)

Graphics Files Format: A connection could not be established allowing the transfer of graphic files (i.e., .bmp, .jpg, .gif .png, .tif) to the Motorola V300. Images were created via the picture camera. No data was found. (Miss)

Compressed File Archive Formats: N.A. – The Motorola V300 does not support compressed archive files (e.g., .zip, .rar, .exe, .tgz). (NA)

Misnamed Files: Misnamed file (e.g., .txt file renamed with a .dll extension) data content was not found when sent via email. The filename and subject line of the email were found and reported. (Miss)

Peripheral Memory Cards: N.A. – The Motorola V300 does not allow for removable media. (NA)

Acquisition Consistency: N.A. – This signature is not a hash of the device data, but the identity of the examiner. (NA)

Cleared Devices: A Hard Reset was performed by selecting, Master Clear and Master Reset options in the settings menu. SMS Messages were recovered. (Above)

Power Loss: The Motorola V300 was repopulated with the above scenarios, then completely drained of all battery power and reacquired. All data was found as reported above. (Above)

Nokia 6610i

The following scenarios were conducted on an unlocked GSM Nokia 6610i. GSM .XRY version 2.5 was used for acquisition.

Connectivity and Retrieval: Proper authentication had to be provided to the password-protected device before contents were successfully acquired. The device contents were successfully acquired with or without the SIM present. Basic subscriber and service provider information was found and reported (i.e., IMEI, IMSI). Memory size is not reported. (Meet)

PIM Applications: All PIM data was found (i.e., Address Book, Calendar) in the corresponding database/folder (i.e., Contacts, Calendar) and reported. Deleted PIM data was not found. (Below)

Dialed/Received Phone Calls: All dialed/received phone calls were found and reported in the Calls folder. Deleted phone calls were not found. (Below)

SMS/MMS Messaging: All active incoming and outgoing SMS messages were found in the SMS Message folder and reported. Deleted SMS/MMS messages were not found. Textual data remnants of received MMS messages were found in the Files folder. Attached MMS data was found in the Picture and Audio folders. (Below)

Internet Messaging: N.A. – The Nokia 6610i does not support email. (NA)

Web Applications: Visited URLs and search engine queries were found and reported. Textual Web content pertaining to URLs were not found. No graphical images of visited sites were found or displayed. (Below)

Text File Formats: Data content associated with text files (i.e., .txt, .doc, .pdf) was found and reported in the Files folder and could be viewed after saved to the forensic workstation. Deleted text files were not found. (Below)

Graphics Files Format: Graphic files (i.e., .bmp, .jpg, .gif, .png, .tif) present on the device were found and reported in the Pictures folder. Deleted graphic files were not found. (Below)

Compressed File Archive Formats: Compressed data file (i.e., .zip, .rar, .exe, .tgz) content was found and reported in the Files folder and could be viewed after saved to the forensic workstation. (Meet)

Misnamed Files: Misnamed files (e.g., .txt file renamed with a .dll extension) were found and reported in the Files folder and could be viewed with the proper application after saved to the forensic workstation. (Meet)

Peripheral Memory Cards: N.A. – The Nokia 6610i does not allow for removable media. (NA)

Acquisition Consistency: N.A. – This signature is not a hash of the device data, but the identity of the examiner. (NA)

Cleared Devices: N.A. – A Hard Reset function is not provided by the phone. (NA)

Power Loss: The Nokia 6610i was repopulated with the above scenarios, then completely drained of all battery power and reacquired. All data was found as reported above. (Above)

Nokia 6200

The following scenarios were conducted on a GSM Nokia 6200. GSM .XRY version 2.5 was used for acquisition.

Connectivity and Retrieval: Proper authentication had to be provided to the password-protected device before contents were successfully acquired. The device contents were successfully acquired with or without the SIM present. Basic subscriber and service provider information was found and reported (i.e., IMEI). Memory size is not reported. (Meet)

PIM Applications: All PIM data was found (i.e., Address Book, Calendar, Tasks) in the corresponding database/folder (i.e., Contacts, Calendar, Notes) and reported. Deleted PIM data was not found. (Below)

Dialed/Received Phone Calls: Dialed/received phone calls were found and reported. Deleted phone calls were not found. (Below)

SMS/MMS Messaging: All active incoming and outgoing SMS text messages were found and reported. Deleted SMS/MMS messages were not found. MMS messages with attachments (i.e., graphics, audio files) were found. (Below)

Internet Messaging: N.A. The Nokia 6200 does not support email. (NA)

Web Applications: Visited URLs, search queries performed, textual Web content or graphical images of visited sites were not found. (Miss)

Text File Formats: Data content associated with text files (i.e., .txt, .doc, .pdf) was found and reported in the Files folder and could be viewed after saved to the forensic workstation. Deleted text files were not found. (Below)

Graphics Files Format: Graphic files (i.e., .bmp, .jpg, .gif, .png, .tif) present on the device were found and reported in the Pictures folder. Deleted graphic files were not found. (Below)

Compressed File Archive Formats: Compressed data file (i.e., .zip, .rar, .exe, .tgz) content was found and reported in the Files folder and could be viewed after saved to the forensic workstation. (Meet)

Misnamed Files: Misnamed files (e.g., .txt file renamed with a .dll extension) were found and reported in the Files folder and could be viewed with the proper application after saved to the forensic workstation. (Meet)

Peripheral Memory Cards: N.A. – The Nokia 6200 does not allow for removable media. (NA)

Acquisition Consistency: N.A. – This signature is not a hash of the device data, but the identity of the examiner. (NA)

Cleared Devices: N.A. – A Hard Reset function is not provided by the phone. (NA)

Power Loss: The Nokia 6200 was repopulated with the above scenarios, then completely drained of all battery power and reacquired. All data was found as reported above. (Above)

Nokia 7610

The following scenarios were conducted on a GSM Nokia 7610 running Symbian OS. GSM .XRY version 2.5 was used for acquisition.

Connectivity and Retrieval: GSM .XRY failed to acquire data contents via a cable interface. Therefore, Bluetooth was used for acquisition. Proper authentication had to be provided to the password-protected device in order to turn Bluetooth on to allow for connectivity with the GSM .XRY unit. The SIM must be present for acquisition, but authentication did not have to be provided. Basic subscriber and service provider information was found and reported (i.e., IMEI, IMSI). Memory size is not reported. (Meet)

PIM Applications: All PIM data was found (i.e., Address Book, Calendar, Tasks) in the corresponding database/folder (i.e., Contacts, Calendar, Notes) and reported. Deleted PIM data was not found. (Below)

Dialed/Received Phone Calls: Dialed/received phone calls were not found. (Miss)

SMS/MMS Messaging: Incoming and outgoing SMS/MMS text messages (i.e., text based content) were not found. (Miss)

Internet Messaging: Data content associated with sent and received email messages, was not found. (Miss)

Web Applications: Visited URLs, search queries performed, textual Web content or graphical images of visited sites were not found. (Miss)

Text File Formats: Data content associated with text files (i.e., .txt) was found and reported in the Notes folder. Additional text file formats (e.g., .pdf, .doc) were not found. Deleted text file were not found. (Below)

Graphics Files Format: Graphic files (e.g., .bmp, .jpg, .gif, .png, .tif) present on the device were found and reported in the Pictures folder. Deleted graphic files were not found. (Below)

Compressed File Archive Formats: Compressed data file (i.e., .zip, .rar, .exe, .tgz) content was not found. (Miss)

Misnamed Files: Misnamed files (e.g., .txt file renamed with a .dll extension) were found and reported in the Notes folder. (Meet)

Peripheral Memory Cards: Data residing on a 64 MB MMC populated with various files (i.e., text, graphics, audio, misnamed files) were found and reported. Deleted data was not found. (Below)

Acquisition Consistency: N.A. – This signature is not a hash of the device data, but the identity of the examiner. (NA)

Cleared Devices: A Hard Reset was performed by entering *#7370# followed by the call key. No internal phone memory data was found. Data populated onto the MMC card was found and reported. (Meet)

Power Loss: The Nokia 7610 was repopulated with the above scenarios, then completely drained of all battery power and reacquired. All data was found as reported above. (Above)

Appendix E: Oxygen Phone Manager Results

The scenarios were performed on a desktop workstation running Windows XP SP2. Oxygen Phone Manager version 2.6 was used to acquire data from the following cell phones: Nokia 3390, Nokia 6200, Nokia 6610i and Nokia 7610.

Nokia 3390

The following scenarios were conducted on a Nokia 3390 GSM phone, using an AT-ETSI interface (i.e., AT-ETSI phone protocol plug-in, ETSI-AT phone data conversion). OPM II Version 2.6 was used to acquire data from the Nokia 3390.

Connectivity and Retrieval: The password-protected device contents were successfully acquired with or without the SIM present without having to provide proper authentication. Basic subscriber and service provider information (i.e., IMEI), and memory size was found and reported. (Meet)

PIM Applications: All active phone book entries and speed dial numbers were found and reported. Additional PIM Entries were not found. Deleted PIM data was not found. (Below)

Dialed/Received Phone Calls: All dialed/received phone calls were found and reported. Deleted phone calls were not found. (Below)

SMS/MMS Messaging: All active incoming and outgoing SMS messages were found and reported. MMS messages and deleted SMS/MMS messages were not found. (Below)

Internet Messaging: N.A. – The Nokia 3390 does not support email. (NA)

Web Applications: The Nokia 3390 does not support browsing the Web but allows for Instant Messaging. No data was found. (Miss)

Text File Formats: N.A. – The Nokia 3390 does not support text files (e.g., .txt, .doc, .pdf). (NA)

Graphics Files Format: N.A. – The Nokia 3390 does not support graphic files (e.g., .bmp, .jpg, .gif, .png, .tif). (NA)

Compressed File Archive Formats: N.A. – The Nokia 3390 does not support compressed archive files (e.g., .zip, .rar, .exe, .tgz). (NA)

Misnamed Files: N.A. – The Nokia 3390 does not support misnamed files (e.g., .txt file renamed with a .dll extension). (NA)

Peripheral Memory Cards: N.A. – The Nokia 3390 does not allow for removable media. (NA)

Acquisition Consistency: N.A. – OPM II does not provide an internal hashing algorithm for individual files or overall phone acquisition. (NA)

Cleared Devices: N.A. – A Hard Reset function is not provided by the phone. (NA)

Power Loss: The Nokia 3390 was repopulated with the above scenarios, then completely drained of all battery power and reacquired. All data was found as reported above. (Above)

Nokia 6610i

The following scenarios were conducted on an unlocked Nokia 6610i GSM phone, using an AT-ETSI interface (i.e., AT-ETSI phone protocol plug-in, ETSI-AT phone data conversion).

Connectivity and Retrieval: The password-protected device contents were successfully acquired with or without the SIM present without having to provide proper authentication. Basic subscriber, service provider information (i.e., IMEI) and memory size was found and reported. (Meet)

PIM Applications: All PIM data was found (i.e., Address Book, Calendar, Tasks) and reported. Deleted PIM data was not found. (Below)

Dialed/Received Phone Calls: All dialed/received phone calls were found and reported. Deleted phone calls were not found. (Below)

SMS/MMS Messaging: All active incoming and outgoing SMS/MMS messages were found and reported. Deleted SMS/MMS messages were not found. (Below)

Internet Messaging: N.A. – The Nokia 6610i does not support email. (NA)

Web Applications: Visited URLs, search queries performed, textual Web content or graphical images of visited sites were not found. (Miss)

Text File Formats: Data content associated with text files (i.e., .txt, .doc, .pdf) was found and reported and could be viewed after saved to the forensic workstation. Deleted text files were not found. (Below)

Graphics Files Format: Graphic files (i.e., .bmp, .jpg, .gif, .png, .tif) present on the device were found and reported. Deleted graphic files were not found. (Below)

Compressed File Archive Formats: Compressed data file (i.e., .zip, .rar, .exe, .tgz) content was found and reported and could be viewed after saved to the forensic workstation. (Meet)

Misnamed Files: Misnamed files (e.g., .txt file renamed with a .dll extension) were found and reported and could be viewed with the proper application after saved to the forensic workstation. (Meet)

Peripheral Memory Cards: N.A. – The Nokia 6610i does not allow for removable media. (NA)

Acquisition Consistency: N.A. – OPM II does not provide an internal hashing algorithm for individual files or overall phone acquisition. (NA)

Cleared Devices: N.A. – A Hard Reset function is not provided by the phone. (NA)

Power Loss: The Nokia 6610i was repopulated with the above scenarios, then completely drained of all battery power and reacquired. All data was found as reported above. (Above)

Nokia 6200

The following scenarios were conducted on a Nokia 6200 GSM phone, using an AT-ETSI interface (i.e., AT-ETSI phone protocol plug-in, ETSI-AT phone data conversion).

Connectivity and Retrieval: The password-protected device contents were successfully acquired with or without the SIM present providing proper authentication. Basic subscriber, service provider information (i.e., IMEI) and memory size was found and reported. (Meet)

PIM Applications: The following PIM data was found (i.e., Address Book, Calendar, Tasks) and reported. Deleted PIM data was not found. (Below)

Dialed/Received Phone Calls: All dialed/received phone calls were found and reported. Deleted phone calls were not found. (Below)

SMS/MMS Messaging: All active incoming and outgoing SMS/MMS messages were found and reported. Deleted SMS/MMS messages were not found. (Below)

Internet Messaging: N.A. – The Nokia 6200 does not support email. (NA)

Web Applications: Visited URLs, search queries performed, textual Web content or graphical images of visited sites were not found. (Miss)

Text File Formats: Data content associated with text files (i.e., .txt, .doc, .pdf) was found and reported and could be viewed after saved to the forensic workstation. Deleted text files were not found. (Below)

Graphics Files Format: Graphic files (i.e., .bmp, .jpg, .gif, .png, .tif) present on the device were found and reported. Deleted graphic files were not found. (Below)

Compressed File Archive Formats: Compressed data file (i.e., .zip, .rar, .exe, .tgz) content was found and reported and could be viewed after saved to the forensic workstation. (Meet)

Misnamed Files: Misnamed files (e.g., .txt file renamed with a .dll extension) were found and reported and could be viewed with the proper application after saved to the forensic workstation. (Meet)

Peripheral Memory Cards: N.A. – The Nokia 6200 does not allow for removable media. (NA)

Acquisition Consistency: N.A. – OPM II does not provide an internal hashing algorithm for individual files or overall phone acquisition. (NA)

Cleared Devices: N.A. – A Hard Reset function is not provided by the phone. (NA)

Power Loss: The Nokia 6200 was repopulated with the above scenarios, then completely drained of all battery power and reacquired. All data was found as reported above. (Above)

Nokia 7610

The following scenarios were conducted on a Nokia 7610 GSM phone, using a Bluetooth connection.

Connectivity and Retrieval: Proper authentication had to be provided to the password-protected device in order to turn Bluetooth on to allow for connectivity. The SIM must be present for acquisition and authentication must be provided. Basic subscriber and service provider information (i.e., IMEI) and memory size was found and reported. (Meet)

PIM Applications: All PIM data was found (i.e., Address Book, Calendar, Tasks) and reported. Deleted PIM data was not found. (Below)

Dialed/Received Phone Calls: Dialed/received phone calls were not found. (Miss)

SMS/MMS Messaging: All active incoming and outgoing SMS/MMS messages were found and reported. Deleted SMS/MMS messages were not found. (Below)

Internet Messaging: Incoming/outgoing emails were found and reported. Deleted emails were not found. (Below)

Web Applications: Visited URLs, search queries performed, textual Web content or graphical images of visited sites were not found. (Miss)

Text File Formats: Data content associated with text files (i.e., .txt, .doc, .pdf) was not found. (Miss)

Graphics Files Format: Graphic files (i.e., .bmp, .jpg, .gif, .png, .tif) present on the device were found and reported. Deleted graphic files were not found. (Below)

Compressed File Archive Formats: Compressed data file (i.e., .zip, .rar, .exe, .tgz) content was not found. (Miss)

Misnamed Files: Misnamed files (e.g., .txt file renamed with a .dll extension) were not found. (Miss)

Peripheral Memory Cards: Data residing on a 64 MB MMC populated with various files (i.e., text, graphics, audio, misnamed files) were found and reported. Deleted data was not found. (Below)

Acquisition Consistency: N.A. OPM II does not provide an internal hashing algorithm for individual files or overall phone acquisition. (NA)

Cleared Devices: A Hard Reset was performed by entering *#7370# followed by the call key. No internal phone memory data was found. Data populated onto the MMC card was found and reported. (Meet)

Power Loss: The Nokia 7610 was repopulated with the above scenarios, then completely drained of all battery power and reacquired. All data was found as reported above. (Above)

Appendix F: MOBILedit! Results

The scenarios were performed on a Forensic Recovery of Evidence Device (FRED) running Windows XP SP2. MOBILedit! version 1.93 was used to acquire data from the following cell phone: Motorola V66, Nokia 6610i, Motorola V300, Sony Ericsson T68i and a Motorola c333.

Ericsson T68i

The following scenarios were conducted on a Sony Ericsson T68i.

Connectivity and Retrieval: Data-Pilot's Susteen data cable for Sony Ericsson phones was used in order to establish connectivity with Mobiledit!. Proper authentication had to be provided to the password-protected device and the SIM card had to be inserted before contents were successfully acquired. Basic subscriber and service provider information was found and reported (i.e., IMEI, IMSI). Memory size is not reported. (Meet)

PIM Applications: PIM data was not found (i.e., Address Book, Calendar, Tasks). (Miss)

Dialed/Received Phone Calls: All dialed/received phone calls were found and reported in the Missed calls, Last Numbers Dialed, Received calls folders. Deleted phone calls were not found. (Below)

SMS/MMS Messaging: All active incoming/outgoing SMS messages were found in the Inbox folder and reported. Deleted SMS messages were not found. MMS Messages and attachments (i.e., graphics, sound bytes) were not found or reported. (Below)

Internet Messaging: Data content associated with sent and received email messages, was not found. (Miss)

Web Applications: Visited URLs, search queries performed, textual Web content or graphical images of visited sites were not found. (Miss)

Text File Formats: N.A. – The Sony Ericsson T68i does not support text files (e.g., .txt, .doc, .pdf). (NA)

Graphics Files Format: Supported graphic files (i.e., .jpg, .gif) present on the device were not found. (Miss)

Compressed File Archive Formats: N.A. – The Sony Ericsson T68i does not support compressed archive files (e.g., .zip, .rar, .exe, .tgz). (NA)

Misnamed Files: N.A. – The Sony Ericsson T68i does not support misnamed files (e.g., .txt file renamed with a .dll extension). (NA)

Peripheral Memory Cards: N.A. – The Sony Ericsson T68i does not allow for removable media. (NA)

Acquisition Consistency: N.A. – MOBILedit! does not provide an internal hashing algorithm for individual files or overall phone acquisition. (NA)

Cleared Devices: A Hard Reset was performed by selecting, Reset All, in the settings menu. No data was found. (Meet)

Power Loss: The Sony Ericsson T68i was repopulated with the above scenarios, then completely drained of all battery power and reacquired. All data was found as reported above. (Above)

Motorola C333

The following scenarios were conducted on an unlocked GSM Motorola C333.

Connectivity and Retrieval: The device contents were successfully acquired with or without the SIM. Basic subscriber and service provider information was found and reported (i.e., IMEI, IMSI). Memory size is not reported. (Meet)

PIM Applications: All active phone book entries were found and reported. Additional PIM Entries were not found. Deleted PIM data was not found. (Below)

Dialed/Received Phone Calls: All dialed/received phone calls were found and reported in the Last Numbers Dialed, received call folders. Deleted phone calls were not found. (Below)

SMS/MMS Messaging: No incoming or outgoing SMS/MMS messages were found. (Miss)

Internet Messaging: N.A. – The Motorola C333 does not support email. (NA)

Web Applications: N.A. – Internet connectivity was unable to be established. (NA)

Text File Formats: N.A. – The Motorola C333 does not support text files (e.g., .txt, .doc, .pdf). (NA)

Graphics Files Format: N.A. – The Motorola C333 does not support graphic files (e.g., .bmp, .jpg, .gif, .png, .tif). (NA)

Compressed File Archive Formats: N.A. – The Motorola C333 does not support compressed archive files (e.g., .zip, .rar, .exe, .tgz). (NA)

Misnamed Files: N.A. – The Motorola C333 does not support misnamed files (e.g., .txt file renamed with a .dll extension). (NA)

Peripheral Memory Cards: N.A. – The Motorola C333 does not allow for removable media. (NA)

Acquisition Consistency: N.A. – MOBILedit! does not provide an internal hashing algorithm for individual files or overall phone acquisition. (NA)

Cleared Devices: A Hard Reset was performed by selecting the, Master Reset, in the settings menu. Incoming SMS messages were found and reported, no other data was found. (Above)

Power Loss: The Motorola C333 was repopulated with the above scenarios, then completely drained of all battery power and reacquired. All data was found as reported above. (Above)

Motorola V66

The following scenarios were conducted on an unlocked GSM Motorola V.series 66.

Connectivity and Retrieval: The password-protected device contents were successfully acquired with or without the SIM present without having to provide proper authentication. Basic subscriber and service provider information was found and reported (i.e., IMEI, IMSI). Memory size is not reported. (Meet)

PIM Applications: All active phone book entries were found and reported. Additional PIM Entries were not found. Deleted PIM data was not found. (Below)

Dialed/Received Phone Calls: All dialed/received phone calls were found and reported in the Last Numbers Dialed, received call folders. Deleted phone calls were not found. (Below)

SMS/MMS Messaging: No incoming or outgoing SMS messages were found. MMS Messaging is not supported. (Miss)

Internet Messaging: N.A. – The Motorola V66 does not support email. (NA)

Web Applications: N.A. – Internet connectivity was unable to be established. (NA)

Text File Formats: N.A. – The Motorola V66 does not support text files (e.g., .txt, .doc, .pdf). (NA)

Graphics Files Format: N.A. – The Motorola V66 does not support graphic files (e.g., .bmp, .jpg, .gif, .png, .tif). (NA)

Compressed File Archive Formats: N.A. – The Motorola V66 does not support compressed archive files (e.g., .zip, .rar, .exe, .tgz). (NA)

Misnamed Files: N.A. – The Motorola V66 does not support misnamed files (e.g., .txt file renamed with a .dll extension). (NA)

Peripheral Memory Cards: N.A. – The Motorola V66 does not allow for removable media. (NA)

Acquisition Consistency: N.A. – MOBILedit! does not provide an internal hashing algorithm for individual files or overall phone acquisition. (NA)

Cleared Devices: A Hard Reset was performed by selecting, Master Reset, in the settings menu. No data was found. (Meet)

Power Loss: The Motorola V66 was repopulated with the above scenarios, then completely drained of all battery power and reacquired. All data was found as reported above. (Above)

Motorola V300

The following scenarios were conducted on a Pay-As-You-Go GSM Motorola V300.

Connectivity and Retrieval: Proper authentication had to be provided to the password-protected device and the SIM card had to be inserted before contents were successfully acquired. Basic subscriber and service provider information was found and reported (i.e., IMEI, IMSI). Memory size is not reported. (Meet)

PIM Applications: PIM data related to phonebook entries was found in the Phonebook folder. No other PIM data was found. Deleted PIM data was not found. (Below)

Dialed/Received Phone Calls: All dialed/received phone calls were found and reported in the Missed calls, Last Numbers Dialed, Received calls folders. Deleted phone calls were not found. (Below)

SMS/MMS Messaging: All active incoming and outgoing SMS/MMS messages were found in the inbox folder and reported. Deleted SMS/MMS messages were not found. MMS graphic file attachments were found in the Files->picture folder. (Below)

Internet Messaging: Incoming/outgoing emails were found and reported in the inbox. Chat logs and deleted emails were not found. (Below)

Web Applications: Visited URLs, search queries performed, textual Web content or graphical images of visited sites were not found. (Miss)

Text File Formats: Data content associated with text files (e.g., .txt, .doc, .pdf) was not found. (Miss)

Graphics Files Format: A connection could not be established allowing the transfer of graphic files (i.e., .bmp, .jpg, .gif .png, .tif) to the Motorola V300. Images were created via the picture camera. No data was found. (Miss)

Compressed File Archive Formats: N.A. – The Motorola V300 does not support compressed archive files (e.g., .zip, .rar, .exe, .tgz). (NA)

Misnamed Files: N.A. – The Motorola V300 does not support misnamed files (e.g., .txt file renamed with a .dll extension). (NA)

Peripheral Memory Cards: N.A. – The Motorola V300 does not allow for removable media. (NA)

Acquisition Consistency: N.A. – MOBILedit! does not provide an internal hashing algorithm for individual files or overall phone acquisition. (NA)

Cleared Devices: A Hard Reset was performed by selecting, Master Reset, in the settings menu. SMS Messages were recovered. (Above)

Power Loss: The Motorola V300 was repopulated with the above scenarios, then completely drained of all battery power and reacquired. All data was found as reported above. (Above)

Nokia 6610i

The following scenarios were conducted on an unlocked GSM Nokia 6610i.

Connectivity and Retrieval: Proper authentication had to be provided to the password-protected device and the SIM card had to be inserted before contents were successfully acquired. Basic subscriber and service provider information was found and reported (i.e., IMEI). Memory size is not reported. (Meet)

PIM Applications: Contacts added to the phone were found and reported in the Phonebook folder. Entries created in the Calendar and To-Do list were not found. Deleted PIM data was not found. (Below)

Dialed/Received Phone Calls: All dialed/received phone calls were found and reported in the Last Numbers Dialed, Received Calls folder. Deleted phone calls were not found. (Below)

SMS/MMS Messaging: All active incoming and outgoing SMS messages were found in the inbox and sent items folder and reported. Active MMS messages and deleted SMS messages were not found. (Below)

Internet Messaging: N.A. – The Nokia 6610i does not support email. (NA)

Web Applications: Visited URLs, search queries performed, textual Web content or graphical images of visited sites were not found. (Miss)

Text File Formats: Data content associated with text files (i.e., .txt, .doc, .pdf) was found and reported in the Files folder. Deleted text files were not found. (Below)

Graphics Files Format: Graphic files (i.e., .bmp, .jpg, .gif, .png, .tif) present on the device were found and reported in the Files and Graphics folders. Deleted graphic files were not found. (Below)

Compressed File Archive Formats: Compressed data file (i.e., .zip, .rar, .exe, .tgz) content was found and reported in the Files folder. (Meet)

Misnamed Files: Misnamed files (e.g., .txt file renamed with a .dll extension) were found and reported in the Files folder. (Meet)

142

Peripheral Memory Cards: N.A. – The Nokia 6610i does not allow for removable media. (NA)

Acquisition Consistency: N.A. – MOBILedit! does not provide an internal hashing algorithm for individual files or overall phone acquisition. (NA)

Cleared Devices: N.A. – A Hard Reset function is not provided by the phone. (NA)

Power Loss: The Nokia 6610i was repopulated with the above scenarios, then completely drained of all battery power and reacquired. All data was found as reported above. (Above)

Appendix G: BitPIM Results

The scenarios were performed on a Forensic Recovery of Evidence Device (FRED) running Windows XP SP2. BitPIM version 0.7.28 was used to acquire data from the following CDMA cell phones: Audiovox 8910 and a Sanyo 8200.

Audiovox 8910

The following scenarios were conducted on a pre-paid CMDA Audiovox 8910.

Connectivity and Retrieval: The password-protected device contents were successfully acquired. Basic subscriber and service provider information was not found. Memory size is not reported. (Meet)

PIM Applications: All active and remnants of deleted PIM data was found and reported in the Filesystem section. (Meet)

Dialed/Received Phone Calls: All active and deleted dialed/received phone calls were found and reported in the Filesystem section. (Meet)

SMS/MMS Messaging: All active incoming and outgoing SMS messages were found and reported in the Filesystem section. MMS messages and deleted messages were not found. (Below)

Internet Messaging: N.A. – The Audiovox 8910 does not support email. (NA)

Web Applications: Visited URLs were found and reported. Search engine queries, textual Web content pertaining to visited URLs or graphical images of visited sites, were not found. (Below)

Text File Formats: N.A. – The Audiovox 8910 does not support text files (e.g., .txt, .doc, .pdf). (NA)

Graphics Files Format: A connection could not be established allowing the transfer of graphic files (i.e., .bmp, .jpg, .gif .png, .tif) to the Audiovox 8910. Images were created via the picture camera. Graphic files present on the device were found and reported in the Filesystem section. Images were exported and viewed with a third party application. Deleted graphic files were not found. (Below)

Compressed File Archive Formats: N.A. – The Audiovox 8910 does not support compressed archive files (e.g., .zip, .rar, .exe, .tgz). (NA)

Misnamed Files: N.A. – The Audiovox 8910 does not support misnamed files (e.g., .txt file renamed with a .dll extension). (NA)

Peripheral Memory Cards: N.A. – The Audiovox 8910 does not allow for removable media. (NA)

Acquisition Consistency: N.A. – BitPIM does not provide an internal hashing algorithm for individual files or overall phone acquisition. (NA)

Cleared Devices: N.A. – A Hard Reset function is not provided by the phone. (NA)

Power Loss: The Audiovox 8910 was repopulated with the above scenarios, then completely drained of all battery power and reacquired. All data was found as reported above. (Above)

Sanyo PM-8200

The following scenarios were conducted on a CMDA Sanyo 8200.

Connectivity and Retrieval: The password-protected device contents were successfully acquired. Basic subscriber and service provider information was not found. Memory size is not reported. (Meet)

PIM Applications: All active PIM data was found and reported in the corresponding folder (e.g., Phonebook, Calendar). Tasks entries were found and reported in the Filesystem section. Deleted contact PIM data was found and reported in the Filesystem section. Data remnants of deleted Calendar and Tasks entries were found and reported in the Filesystem section. (Meet)

Dialed/Received Phone Calls: All active and deleted dialed/received phone calls were found and reported in the Filesystem section. (Meet)

SMS/MMS Messaging: All active incoming and outgoing SMS and outgoing MMS messages (textual content) were found and reported in the Filesystem section. Active MMS attachments excluding sound bytes were found and reported in the Wallpaper section. Deleted outgoing SMS/MMS messages (textual content) were found and reported in the Filesystem section. Incoming MMS messages (i.e., textual content) were not found. (Meet)

Internet Messaging: Data content associated with sent and received email messages, was not found. (Miss)

Web Applications: Visited URLs were found and reported. Search engine queries, textual Web content pertaining to URLs or graphical images of visited sites were not found. (Below)

Text File Formats: N.A. – The Sanyo PM-8200 does not support text files (e.g., .txt, .doc, .pdf). (NA)

Graphics Files Format: A connection could not be established allowing the transfer of graphic files (i.e., .bmp, .jpg, .gif .png, .tif) to the Sanyo PM-8200. Images were created via the picture camera. Graphic files present on the device were found and reported in the Wallpaper section. Deleted graphic files were not found. (Below)

Compressed File Archive Formats: N.A. – The Sanyo PM-8200 does not support compressed archive files (e.g., .zip, .rar, .exe, .tgz). (NA)

Misnamed Files: N.A. – The Sanyo PM-8200 does not support misnamed files (e.g., `.txt` file renamed with a `.dll` extension). (NA)

Peripheral Memory Cards: N.A. – The Sanyo PM-8200 does not allow for removable media. (NA)

Acquisition Consistency: N.A. – BitPIM does not provide an internal hashing algorithm for individual files or overall phone acquisition. (NA)

Cleared Devices: A Hard Reset was performed by selecting the, Reset option, in the security menu. All active data is recovered. (Above)

Power Loss: The Sanyo PM-8200 was repopulated with the above scenarios, then completely drained of all battery power and reacquired. All data was found as reported above. (Above)

Appendix H: TULP 2G Results

The scenarios were performed on a forensic workstation running Windows XP SP2. TULP 2G version 1.1.0.2 was used to acquire data from the following cell phones: Audiovox 8910, Ericsson T68i, Sony Ericsson P910a, Motorola C333, Motorola V66, Motorola V300, Nokia 6610i and a Nokia 6200 via a data-link cable. After a successful acquisition, the acquired data contents are stored in an XML file.

Audiovox 8910

The following scenarios were conducted on a pre-paid CMDA Audiovox 8910.

Connectivity and Retrieval: The password-protected device contents were successfully acquired when authentication mechanisms were provided with the SIM present. No data was found. (Meet)

PIM Applications: All active phone book entries were found and reported. Entries created in the Calendar, Memo and Tasks were not found. Deleted PIM data was not found. (Below)

Dialed/Received Phone Calls: All active dialed/received phone calls were found and reported. Deleted phone calls were not found. (Below)

SMS/MMS Messaging: No data was found. (Miss)

Internet Messaging: N.A. – The Audiovox 8910 does not support email. (NA)

Web Applications: Visited URLs, search queries performed, textual Web content or graphical images of visited sites were not found. (Miss)

Text File Formats: N.A. – The Audiovox 8910 does not support text files (e.g., .txt, .doc, .pdf). (NA)

Graphics Files Format: A connection could not be established allowing the transfer of graphic files (i.e., .bmp, .jpg, .gif .png, .tif) to the Audiovox 8910. Images were created by using the picture camera. Graphic files present on the device were not found. (Miss)

Compressed File Archive Formats: N.A. – The Audiovox 8910 does not support compressed archive files (e.g., .zip, .rar, .exe, .tgz). (NA)

Misnamed Files: N.A. – The Audiovox 8910 does not support misnamed files (e.g., .txt file renamed with a .dll extension). (NA)

Peripheral Memory Cards: N.A. – The Audiovox 8910 does not allow for removable media. (NA)

Acquisition Consistency: N.A. – The MD5/SHA1 hashing algorithms are used to determine whether a case file has been damaged, not to verify that the acquired data has not been altered after acquisition. (NA)

Cleared Devices: N.A. – A Hard Reset function is not provided by the phone. (NA)

Power Loss: The Audiovox 8910 was repopulated with the above scenarios, then completely drained of all battery power and reacquired. All data was found as reported above. (Above)

Ericsson T68i

The following scenarios were conducted on an unlocked Sony Ericsson T68i.

Connectivity and Retrieval: The password-protected device contents were successfully acquired when authentication mechanisms were provided with or without the SIM present. Basic subscriber and service provider information was found and reported (i.e., IMEI). Memory size is not reported. (Meet)

PIM Applications: All active phone book entries were found and reported. Entries created in the Calendar, Memo and Tasks were not found. Deleted PIM data was not found. (Below)

Dialed/Received Phone Calls: All dialed/received phone calls were found and reported. Deleted phone calls were not found. (Below)

SMS/MMS Messaging: All active incoming and outgoing SMS messages were found and reported. Deleted SMS messages were not found. MMS Messages and attachments (i.e., graphics, sound bytes) were not found or reported. (Below)

Internet Messaging: Data content associated with sent and received email messages, was not found. (Miss)

Web Applications: Visited URLs, search queries performed, textual Web content or graphical images of visited sites were not found. (Miss)

Text File Formats: N.A. – The Sony Ericsson T68i does not support text files (e.g., .txt, .doc, .pdf). (NA)

Graphics Files Format: Supported graphic files (i.e., .jpg, .gif) present on the device were not found. (Miss)

Compressed File Archive Formats: N.A. – The Sony Ericsson T68i does not support compressed archive files (e.g., .zip, .rar, .exe, .tgz). (NA)

Misnamed Files: N.A. – The Sony Ericsson T68i does not support misnamed files (e.g., .txt file renamed with a .dll extension). (NA)

Peripheral Memory Cards: N.A. – The Sony Ericsson T68i does not allow for removable media. (NA)

Acquisition Consistency: N.A. – The MD5/SHA1 hashing algorithms are used to determine whether a case file has been damaged, not to verify that the acquired data has not been altered after acquisition. (NA)

Cleared Devices: A Hard Reset was performed by selecting, Master Reset, in the settings menu. No data was found. (Meet)

Power Loss: The Sony Ericsson T68i was repopulated with the above scenarios, then completely drained of all battery power and reacquired. All data was found as reported above. (Above)

Sony Ericsson p910a

The following scenarios were conducted on an unlocked Sony Ericsson p910a.

Connectivity and Retrieval: The password-protected device contents were successfully acquired when authentication mechanisms were provided with or without the SIM present. Basic subscriber and service provider information was found and reported (i.e., IMEI). Memory size is not reported. (Meet)

PIM Applications: No PIM data was found. (Miss)

Dialed/Received Phone Calls: Dialed/received phone calls were not found. (Miss)

SMS/MMS Messaging: No incoming or outgoing SMS/MMS messages were found. (Miss)

Internet Messaging: Data content associated with sent and received email messages, was not found. (Miss)

Web Applications: Visited URLs, search queries performed, textual Web content or graphical images of visited sites were not found. (Miss)

Text File Formats: Data content associated with text files (e.g., .txt, .doc, .pdf) was not found. (Miss)

Graphics Files Format: Supported graphic files (i.e., .jpg, .gif) present on the device were not found. (Miss)

Compressed File Archive Formats: Compressed data file (i.e., .zip, .rar, .exe, .tgz) content was not found. (Miss)

Misnamed Files: Misnamed files (e.g., .txt file renamed with a .dll extension) were not found. (Miss)

Peripheral Memory Cards: No data residing on a 128 MB MMC populated with various files (i.e., text, graphics, audio, compressed archive files, misnamed files) was found. (Miss)

Acquisition Consistency: N.A. – The MD5/SHA1 hashing algorithms are used to determine whether a case file has been damaged, not to verify that the acquired data has not been altered after acquisition. (NA)

Cleared Devices: A Hard Reset was performed by selecting, Master Reset, in the settings menu. No data was found. (Meet)

Power Loss: The Sony Ericsson p910a was repopulated with the above scenarios, then completely drained of all battery power and reacquired. No data was found. (Meet)

Motorola C333

The following scenarios were conducted on an unlocked Motorola C333 GSM phone, using an AT-ETSI interface (i.e., AT-ETSI phone protocol plug-in, ETSI-AT phone data conversion).

Connectivity and Retrieval: The password-protected device contents were successfully acquired with SIM present and providing proper authentication. Basic subscriber, service provider information (i.e., IMEI) was not found. Memory size is not reported. (Meet)

PIM Applications: All active phone book entries were found and reported. Additional PIM entries were not found. Deleted PIM data was not found. (Below)

Dialed/Received Phone Calls: All dialed/received phone calls were found and reported. Deleted phone calls were not found. (Below)

SMS/MMS Messaging: No incoming or outgoing SMS/MMS messages were found. (Miss)

Internet Messaging: N.A. – The Motorola C333 does not support email. (NA)

Web Applications: N.A. – Internet connectivity was unable to be established. (NA)

Text File Formats: N.A. – The Motorola C333 does not support test files (e.g., .txt, .doc, .pdf). (NA)

Graphics Files Format: N.A. – The Motorola C333 does not support graphic files (e.g., .bmp, .jpg, .gif, .png, .tif). (NA)

Compressed File Archive Formats: N.A. – The Motorola C333 does not support compressed archive files (e.g., .zip, .rar, .exe, .tgz). (NA)

Misnamed Files: N.A. – The Motorola C333 does not support misnamed files (e.g., .txt file renamed with a .dll extension). (NA)

Peripheral Memory Cards: N.A. – The Motorola C333 does not allow for removable media. (NA)

Acquisition Consistency: N.A. – The MD5/SHA1 hashing algorithms are used to determine whether a case file has been damaged, not to verify that the acquired data has not been altered after acquisition. (NA)

Cleared Devices: A Hard Reset was performed by selecting the, Master Reset option, in the settings menu. No data was found. (Meet)

Power Loss: The Motorola C333 was repopulated with the above scenarios, then completely drained of all battery power and reacquired. All data was found as reported above. (Above)

Motorola V66

The following scenarios were conducted on an unlocked Motorola V.series 66 GSM phone, using an AT-ETSI interface (i.e., AT-ETSI phone protocol plug-in, ETSI-AT phone data conversion).

Connectivity and Retrieval: The password-protected device contents were successfully acquired with or without the SIM present providing proper authentication. Basic subscriber and service provider information was not found. Memory size is not reported. (Meet)

PIM Applications: All active phone book entries were found and reported. Additional PIM entries were not found. Deleted PIM data was not found. (Below)

Dialed/Received Phone Calls: All dialed/received phone calls were found and reported. Deleted phone calls were not found. (Below)

SMS/MMS Messaging: No incoming or outgoing SMS messages were found. MMS Messaging is not supported. (Miss)

Internet Messaging: N.A. – The Motorola V66 does not support email. (NA)

Web Applications: N.A. – Internet connectivity was unable to be established. (NA)

Text File Formats: N.A. – The Motorola V66 does not support text files (e.g., .txt, .doc, .pdf). (NA)

Graphics Files Format: N.A. – The Motorola V66 does not support graphic files (e.g., .bmp, .jpg, .gif, .png, .tif). (NA)

Compressed File Archive Formats: N.A. – The Motorola V66 does not support compressed archive files (e.g., .zip, .rar, .exe, .tgz). (NA)

Misnamed Files: N.A. – The Motorola V66 does not support misnamed files (e.g., .txt file renamed with a .dll extension). (NA)

Peripheral Memory Cards: N.A. – The Motorola V66 does not allow for removable media. (NA)

Acquisition Consistency: N.A. – The MD5/SHA1 hashing algorithms are used to determine whether a case file has been damaged, not to verify that the acquired data has not been altered after acquisition. (NA)

Cleared Devices: A Hard Reset was performed by selecting the, Master Clear option, in the settings menu. No data was found. (Meet)

Power Loss: The Motorola V66 was repopulated with the above scenarios, then completely drained of all battery power and reacquired. All data was found as reported above. (Above)

Motorola V300

The following scenarios were conducted on an unlocked Motorola V300 GSM phone, using an AT-ETSI interface (i.e., AT-ETSI phone protocol plug-in, ETSI-AT phone data conversion).

Connectivity and Retrieval: Proper authentication had to be provided to the password-protected device and the SIM card had to be inserted before contents were successfully acquired. Basic subscriber and service provider information (e.g., Equipment Identifier, IMEI) were not found. Memory size and the ICCID are not reported. (Meet)

PIM Applications: All active phone book entries were found and reported. Additional PIM entries were not found. Deleted PIM data was not found. (Below)

Dialed/Received Phone Calls: All dialed/received phone calls were found and reported. Deleted phone calls were not found. (Below)

SMS/MMS Messaging: All active incoming SMS messages were found and reported. Outgoing/deleted SMS and MMS messages were not found. (Below)

Internet Messaging: Incoming/outgoing emails were found and reported. Deleted messages were not found. (Below)

Web Applications: Visited URLs, search queries performed, textual Web content or graphical images of visited sites were not found. (Miss)

Text File Formats: Data content associated with text files (e.g., .txt, .doc, .pdf) was not found. (Miss)

Graphics Files Format: A connection could not be established allowing the transfer of graphic files (e.g., .bmp, .jpg, .gif .png, .tif) to the Motorola V300. Images were created by using the picture camera. No data was found. (Miss).

Compressed File Archive Formats: N.A. – The Motorola V300 does not support compressed archive files (e.g., .zip, .rar, .exe, .tgz). (NA)

152

Misnamed Files: N.A. – The Motorola V300 does not support misnamed files (e.g., .txt file renamed with a .dll extension). (NA)

Peripheral Memory Cards: N.A. – The Motorola V300 does not allow for removable media. (NA)

Acquisition Consistency: N.A. – The MD5/SHA1 hashing algorithms are used to determine whether a case file has been damaged, not to verify that the acquired data has not been altered after acquisition. (NA)

Cleared Devices: A Hard Reset was performed by selecting the, Master Reset option, in the settings menu. No data was found. (Meet)

Power Loss: The Motorola V300 was repopulated with the above scenarios, then completely drained of all battery power and reacquired. All data was found as reported above. (Above)

Nokia 6610i

The following scenarios were conducted on an unlocked Nokia 6610i GSM phone, using an AT-ETSI interface (i.e., AT-ETSI phone protocol plug-in, ETSI-AT phone data conversion).

Connectivity and Retrieval: The password-protected device contents were successfully acquired with or without the SIM present without providing authentication. Basic subscriber and service provider information was found and reported (i.e., IMEI, ICCID). Memory size is not reported. (Meet)

PIM Applications: Active phonebook entries were found and reported. No other PIM data was found. (Below)

Dialed/Received Phone Calls: All dialed/received phone calls were found and reported. Deleted phone calls were not found. (Below)

SMS/MMS Messaging: All active incoming and outgoing SMS messages were found and reported. Active MMS and deleted SMS/MMS messages were not found. (Below)

Internet Messaging: N.A. – The Nokia 6610i does not support email. (NA)

Web Applications: Visited URLs, search queries performed, textual Web content or graphical images of visited sites were not found. (Miss)

Text File Formats: Data content associated with text files (i.e., .txt, .doc, .pdf) was not found. (Miss)

Graphics Files Format: Graphic files (e.g., .bmp, .jpg, .gif, .png, .tif) were not found. (Miss)

Compressed File Archive Formats: Compressed data file (i.e., .zip, .rar, .exe, .tgz) content was not found. (Miss)

Misnamed Files: Misnamed files (e.g., .txt file renamed with a .dll extension) were not found. (Miss)

Peripheral Memory Cards: N.A. – The Nokia 6610i does not allow for removable media. (NA)

Acquisition Consistency: N.A. – The MD5/SHA1 hashing algorithms are used to determine whether a case file has been damaged, not to verify that the acquired data has not been altered after acquisition. (NA)

Cleared Devices: N.A. – A Hard Reset function is not provided by the phone. (NA)

Power Loss: The Nokia 6610i was repopulated with the above scenarios, then completely drained of all battery power and reacquired. All data was found as reported above. (Above)

Nokia 6200

The following scenarios were conducted on a Nokia 6200 GSM phone, using an AT-ETSI interface (i.e., AT-ETSI phone protocol plug-in, ETSI-AT phone data conversion).

Connectivity and Retrieval: The password-protected device contents were successfully acquired with SIM present and providing proper authentication. Basic subscriber, service provider information (i.e., IMEI) was found and reported. (Meet)

PIM Applications: Active phonebook entries were found and reported. No other PIM data was found. (Below)

Dialed/Received Phone Calls: All dialed/received phone calls were found and reported. Deleted phone calls were not found. (Below)

SMS/MMS Messaging: All active incoming and outgoing SMS messages were found and reported. Active MMS and deleted SMS/MMS messages were not found. (Below)

Internet Messaging: N.A. – The Nokia 6200 does not support email. (NA)

Web Applications: Visited URLs, search queries performed, textual Web content or graphical images of visited sites were not found. (Miss)

Text File Formats: Data content associated with text files (i.e., .txt, .doc, .pdf) was not found. (Miss)

Graphics Files Format: Graphic files (e.g., .bmp, .jpg, .gif, .png, .tif) were not found. (Miss)

Compressed File Archive Formats: Compressed data file (i.e., .zip, .rar, .exe, .tgz) content was not found. (Miss)

Misnamed Files: Misnamed files (e.g., .txt file renamed with a .dll extension) were not found. (Miss)

Peripheral Memory Cards: N.A. – The Nokia 6200 does not allow for removable media. (NA)

Acquisition Consistency: N.A. – The MD5/SHA1 hashing algorithms are used to determine whether a case file has been damaged, not to verify that the acquired data has not been altered after acquisition. (NA)

Cleared Devices: N.A. – A Hard Reset function is not provided by the phone. (NA)

Power Loss: The Nokia 6200 was repopulated with the above scenarios, then completely drained of all battery power and reacquired. All data was found as reported above. (Above)

Appendix I: TULP 2G – External SIM Results

The scenarios were performed on a Forensic Recovery of Evidence Device (FRED) running Windows XP SP2. TULP 2G version 1.1.0.2 was used with a GemPlus PC/SC Reader to acquire data from a populated SIM.

SIM 5343

The following scenarios were conducted using a T-Mobile SIM. Outgoing SMS messages were not populated, and Service Provider Name (SPN) was not allocated. No FPLMN entries were registered.

Basic Data: The following data was found and reported: IMSI, ICCID, Language Preference (LP), Abbreviated Dialing Numbers (ADN), Last Numbers Dialed (LND) and active/deleted incoming SMS messages. (Meet)

Location Data: The following LOCI data was found and reported: TMSI, partial LAI (i.e., LAC), and Status. The following data was not found: GPRSLOCI and the MCC/MNC portions of LAI within LOCI. (Below)

EMS Data: Active/deleted incoming EMS Messages that exceed 160 characters or contain a small, 16x16 pixel, embedded picture were found and reported. (Meet)

Foreign Language Data: ADN entries and SMS messages containing French and Asian language characters were found and reported. (Meet)

SIM 8778

The following scenarios were conducted using a Cingular SIM. Outgoing SMS Messages were not populated, and Service Provider Name (SPN) was allocated but not activated. No FPLMN entries were registered.

Basic Data: The following data was found and reported: IMSI, ICCID, Language Preference (LP), Abbreviated Dialing Numbers (ADN), Last Numbers Dialed (LND) and active/deleted SMS messages. A blank outgoing message was incorrectly reported as containing 189 extraneous characters. (Meet)

Location Data: The following LOCI data was found and reported: TMSI, partial LAI (i.e., LAC), and Status. The following data was not found: GPRSLOCI and the MCC/MNC portions of LAI within LOCI. (Below)

EMS Data: Active/deleted incoming EMS Messages that exceed 160 characters or contain a small, 16x16 pixel, embedded picture were found and reported. (Meet)

Foreign Language Data: ADN entries and SMS messages containing French and Asian language characters were found and reported. (Meet)

SIM 1144

The following scenarios were conducted using an AT&T SIM. Service Provider Name (SPN) was not allocated. No FPLMN entries were registered.

Basic Data: The following data was found and reported: IMSI, ICCID, Language Preference (LP), Abbreviated Dialing Numbers (ADN), Last Numbers Dialed (LND) and active/deleted SMS messages. (Meet)

Location Data: The following LOCI data was found and reported: TMSI, partial LAI (i.e., LAC), and Status. The following data was not found: GPRSLOCI and the MCC/MNC portions of LAI within LOCI. (Below)

EMS Data: Active/deleted incoming EMS Messages that exceed 160 characters were found and reported. An EMS message containing a large, 32x32 pixel, embedded picture was not found. (Below)

Foreign Language Data: ADN entries and SMS messages containing French and Asian language characters were found and reported. (Meet)

Appendix J: SIMIS Results

The scenarios were performed on a Forensic Recovery of Evidence Device (FRED) running Windows XP SP2. SIMIS version 2.0.13 was used to acquire data from a populated SIM.

SIM 5343

The following scenarios were conducted using a T-Mobile SIM. Outgoing SMS messages were not populated, and Service Provider Name (SPN) was not allocated. No FPLMN entries were registered.

Basic Data: The following data was found and reported: IMSI, ICCID, Language Preference (LP), Abbreviated Dialing Numbers (ADN), Last Numbers Dialed (LND) and active/deleted incoming SMS messages. The reported IMSI value was prepended with length and parity data. Long ADN text entries were truncated to 14 characters in the user interface display, but the full text appeared in the .dmp file. (Meet)

Location Data: All LOCI and LOCIGPRS data was found and reported. (Meet)

EMS Data: Active/deleted incoming EMS Messages that exceed 160 characters were found and reported. An EMS message containing an embedded picture was found, but the image was not correctly decoded and presented, though the text was. (Below)

Foreign Language Data: ADN entries and SMS messages containing French and Asian language characters were found and reported. However, ADN entries and SMS messages containing French characters were not displayed correctly, although the messages appeared correctly the .dmp file. (Below)

SIM 8778

The following scenarios were conducted using a Cingular SIM. Service Provider Name (SPN) was allocated but not activated. No FPLMN entries were registered.

Basic Data: The following data was found and reported: IMSI, ICCID, Language Preference (LP), Abbreviated Dialing Numbers (ADN), Last Numbers Dialed (LND) and active/deleted SMS messages. The reported IMSI value was prepended with length and parity data. Long ADN text entries were truncated to 14 characters in the user interface display, but the full text appeared in the .dmp file. (Meet)

Location Data: All LOCI and LOCIGPRS data was found and reported. However, the MNC portion of the LAI, a three-digit value, was incorrectly decoded in both the user-interface and the .dmp file. (Below)

EMS Data: Active/deleted incoming EMS Messages that exceed 160 characters were found and reported. An EMS message containing an embedded picture was found, but the image was not correctly decoded and presented, though the text was. (Below)

Foreign Language Data: ADN entries and SMS messages containing French and Asian language characters were found and reported. However, ADN entries and SMS messages containing French characters were not displayed correctly, although the messages appeared correctly the .dmp file. (Below)

SIM 1144

The following scenarios were conducted using an AT&T SIM. Service Provider Name (SPN) was not allocated. No FPLMN entries were registered.

Basic Data: The following data was found and reported: IMSI, ICCID, Language Preference (LP), Abbreviated Dialing Numbers (ADN), Last Numbers Dialed (LND) and active/deleted SMS messages. The reported IMSI value was prepended with length and parity data. Long ADN text entries were truncated to 14 characters in the user interface display, but the full text appeared in the .dmp file. (Meet)

Location Data: All LOCI and LOCIGPRS data was found and reported. However, the MNC portion of the LAI, a three-digit value, was incorrectly decoded in both the user-interface and the .dmp file. (Below)

EMS Data: Active/deleted incoming EMS Messages that exceed 160 characters were found and reported. An EMS message containing an embedded picture was found, but neither the text nor the image were correctly decoded and presented. (Below)

Foreign Language Data: ADN entries and SMS messages containing French and Asian language characters were found and reported. However, ADN entries and SMS messages containing French characters were not displayed correctly, although the messages appeared correctly the .dmp file. (Below)

Appendix K: ForensicSIM Results

The scenarios were performed on a Forensic Recovery of Evidence Device (FRED) running Windows XP SP2. The ForensicSIM Terminal was used to duplicate the target SIM card. ForensicSIM Analysis version 1.3.0.0 was used to acquire data from a populated SIM.

SIM 5343

The following scenarios were conducted using a T-Mobile SIM. Outgoing SMS messages were not populated, and Service Provider Name (SPN) was not allocated. No FPLMN entries were registered.

Basic Data: The following data was found and reported: IMSI, ICCID, Language Preference (LP), Abbreviated Dialing Numbers (ADN), Last Numbers Dialed (LND) and active/deleted incoming SMS messages. (Meet)

Location Data: All LOCI and LOCIGPRS data was found and reported. (Meet)

EMS Data: Active/deleted incoming EMS Messages that exceed 160 characters were found and reported. An EMS message containing an embedded picture was found. However, the image was not decoded and presented, though the text was. (Below)

Foreign Language Data: ADN entries and SMS messages containing French and Asian language characters were found and reported. However, the Asian language SMS messages were presented incorrectly. (Below)

SIM 8778

The scenarios above were conducted using a Cingular SIM. The attempt to acquire the Cingular SIM 8778 was unsuccessful. The error thrown reported: "Target invalid, please remove." (MISS)

SIM 1144

The following scenarios were conducted using an AT&T SIM. Service Provider Name (SPN) was not allocated. No FPLMN entries were registered.

Basic Data: The following data was found and reported: IMSI, ICCID, Language Preference (LP), Abbreviated Dialing Numbers (ADN), Last Numbers Dialed (LND) and active/deleted SMS messages. However, the MNC portion of the IMSI, a three-digit value, was incorrectly decoded and reported. (Below)

Location Data: All LOCI and LOCIGPRS data was found and reported. However, the MCC and MNC portions of the LAI and RAI were incorrectly decoded and reported. (Below)

EMS Data: Active/deleted incoming EMS Messages that exceed 160 characters were found and reported. An EMS message containing an embedded picture was found. However, the image was not decoded and presented, though the text was. (Below)

Foreign Language Data: ADN entries and SMS messages containing French and Asian language characters were found and reported. However, the Asian language SMS messages were presented incorrectly. (Below)

Appendix L: Forensic Card Reader Results

The scenarios were performed on a Forensic Recovery of Evidence Device (FRED) running Windows XP SP2. Forensic Card Reader version 1.0.1 was used with the FCR SIM Card Reader to acquire data from a populated SIM.

SIM 5343

The following scenarios were conducted using a T-Mobile SIM. Outgoing SMS messages were not populated, and Service Provider Name (SPN) was not allocated. No FPLMN entries were registered.

Basic Data: The following data was found and reported: IMSI, ICCID, Abbreviated Dialing Numbers (ADN), Last Numbers Dialed (LND) and active incoming SMS messages. The reported IMSI value was prepended with a parity quartet. The following data was not found: Language Preference (LP) and deleted SMS messages. (Below)

Location Data: All LOCI data was found and reported, but the LAI had to be manually decoded for interpretation. No GPRSLOCI data was found. (Below)

EMS Data: Active incoming EMS Messages that exceed 160 characters were found and reported. An EMS message containing an embedded picture was found. However, the image was not correctly decoded and presented, though the text was. Deleted EMS messages were not found. (Below)

Foreign Language Data: ADN entries and SMS messages containing French language characters were found and reported. ADN entries and SMS messages containing Asian language characters were found, but the text characters were not reported. (Below)

SIM 8778

The following scenarios were conducted using a Cingular SIM. Service Provider Name (SPN) was allocated but not activated. No FPLMN entries were registered.

Basic Data: The following data was found and reported: IMSI, ICCID, Abbreviated Dialing Numbers (ADN), Last Numbers Dialed (LND) and active incoming SMS messages. The reported IMSI value was prepended with a parity quartet. The following data was not found: Language Preference (LP) and deleted SMS messages. (Below)

Location Data: All LOCI data was found and reported, but the LAI had to be manually decoded for interpretation. No GPRSLOCI data was found. (Below)

EMS Data: Active incoming EMS Messages that exceed 160 characters were found and reported. An EMS message containing an embedded picture was found. However, the image was not correctly decoded and presented, though the text was. Deleted EMS messages were not found. (Below)

162

Foreign Language Data: ADN entries and SMS messages containing French language characters were found and reported. ADN entries and SMS messages containing Asian language characters were found, but the text characters were not reported. (Below)

SIM 1144

The following scenarios were conducted using an AT&T SIM. Service Provider Name (SPN) was not allocated. No FPLMN entries were registered.

Basic Data: The following data was found and reported: IMSI, ICCID, Abbreviated Dialing Numbers (ADN), Last Numbers Dialed (LND) and active incoming SMS messages. The reported IMSI value was prepended with a parity quartet and outgoing messages were truncated by one character. The following data was not found: Language Preference (LP) and deleted SMS messages. (Below)

Location Data: All LOCI data was found and reported, but the LAI had to be manually decoded for interpretation. No GPRSLOCI data was found. (Below)

EMS Data: Active incoming EMS Messages that exceed 160 characters were found and reported. An EMS message containing an embedded picture was found. However, the image was not correctly decoded and presented, and the text was truncated by one character. Deleted EMS messages were not found. (Below)

Foreign Language Data: ADN entries and SMS messages containing French language characters were found and reported. ADN entries and SMS messages containing Asian language characters were found, but the text characters were not reported. (Below)

Appendix M: SIMCon Results

The scenarios were performed on a Forensic Recovery of Evidence Device (FRED) running Windows XP SP2. SIMCon version 1.1 was used with a GemPlus PC/SC Reader to acquire data from a populated SIM.

SIM 5343

The following scenarios were conducted using a T-Mobile SIM. Outgoing SMS messages were not populated, and Service Provider Name (SPN) was not allocated. No FPLMN entries were registered.

Basic Data: The following data was found and reported: IMSI, ICCID, Language Preference (LP), Abbreviated Dialing Numbers (ADN), Last Numbers Dialed (LND) and active/deleted incoming SMS messages. (Meet)

Location Data: All LOCI and LOCIGPRS data was found and reported. (Meet)

EMS Data: Active/deleted incoming EMS Messages that exceed 160 characters or contain a small, 16x16 pixel, embedded picture were found and reported. (Meet)

Foreign Language Data: ADN entries and SMS messages containing French and Asian language characters were found and reported. (Meet)

SIM 8778

The following scenarios were conducted using a Cingular SIM. Service Provider Name (SPN) was allocated but not activated. No FPLMN entries were registered.

Basic Data: The following data was found and reported: IMSI, ICCID, Language Preference (LP), Abbreviated Dialing Numbers (ADN), Last Numbers Dialed (LND) and active/deleted SMS messages. However, the MNC portion of the IMSI, a three-digit value, was incorrectly translated. (Below)

Location Data: All LOCI and LOCIGPRS data was found and reported. (Meet)

EMS Data: Active/deleted incoming EMS Messages that exceed 160 characters or contain an embedded picture were found and reported. (Meet)

Foreign Language Data: ADN entries and SMS messages containing French and Asian language characters were found and reported. (Meet)

SIM 1144

The following scenarios were conducted using an AT&T SIM. Service Provider Name (SPN) was not allocated. No FPLMN entries were registered.

Basic Data: The following data was found and reported: IMSI, ICCID, Language Preference (LP), Abbreviated Dialing Numbers (ADN), Last Numbers Dialed (LND) and active/deleted

SMS messages. However, the MNC portion of the IMSI, a three-digit value, was incorrectly translated. (Below)

Location Data: All LOCI and LOCIGPRS data was found and reported. (Meet)

EMS Data: Active/deleted incoming EMS Messages that exceed 160 characters or contain an embedded picture were found and reported. (Meet)

Foreign Language Data: ADN entries and SMS messages containing French and Asian language characters were found and reported. (Meet)

Appendix N: Cell Seizure – External SIM Results

The scenarios were performed on a Forensic Recovery of Evidence Device (FRED) running Windows XP SP2. Cell Seizure version 2.0.0.33660 was used with the Cell Seizure SIM Card Reader to acquire data from a populated SIM.

SIM 5343

The following scenarios were conducted using a T-Mobile SIM. Outgoing SMS messages were not populated, and Service Provider Name (SPN) was not allocated. No FPLMN entries were registered.

Basic Data: The following data was found and reported: IMSI, ICCID, Language Preference (LP), Abbreviated Dialing Numbers (ADN), Last Numbers Dialed (LND) and active/deleted incoming SMS messages. The ICCID and LP had to be manually decoded for interpretation. However, maximum length ADN text entries were truncated by one character in the Data view, but not the Binary data view. (Meet)

Location Data: All LOCI and LOCIGPRS data was found and reported, but had to be manually decoded for interpretation. (Meet)

EMS Data: Active/deleted incoming EMS Messages that exceed 160 characters were found and reported. An EMS message containing an embedded picture was found, but neither the text nor the image were correctly decoded and presented. (Below)

Foreign Language Data: SMS messages containing French language characters were found and reported correctly, but French language ADN entries were presented incorrectly in both the user interface and the HTML generated report. SMS messages containing Japanese language characters were found, but reported incorrectly in the user interface and the generated HTML report. The Japanese language ADN entries were presented correctly in the HTML report, but not in the user interface. (Below)

SIM 8778

The scenarios above were conducted using a Cingular SIM. The attempt to acquire the Cingular SIM 8778 was unsuccessful. The error thrown reported: "In this SIM card this type of directory not present." (MISS)

SIM 1144

The following scenarios were conducted using an AT&T SIM. Service Provider Name (SPN) was not allocated. No FPLMN entries were registered.

Basic Data: The following data was found and reported: IMSI, ICCID, Language Preference (LP), Abbreviated Dialing Numbers (ADN), Last Numbers Dialed (LND) and active/deleted SMS messages. However, long ADN text entries were truncated by one character in the Data view, but not the Binary data view. (Meet)

Location Data: All LOCI and LOCIGPRS data was found and reported, but had to be manually decoded for interpretation. (Meet)

EMS Data: Active/deleted incoming EMS Messages that exceed 160 characters were found and reported. An EMS message containing an embedded picture was found, but neither the text nor the image were correctly decoded and presented. (Below)

Foreign Language Data: SMS messages containing French language characters were found and reported correctly, but French language ADN entries were presented incorrectly in both the user interface and the HTML generated report. SMS messages containing Japanese language characters were found, but reported incorrectly in the user interface and the generated HTML report. The Japanese language ADN entries were presented correctly in the HTML report, but not in the user interface. (Below)

Appendix O: GSM .XRY – External SIM Results

The scenarios were performed on a Forensic Recovery of Evidence Device (FRED) running Windows XP SP2. GSM .XRY version 2.5 was used with the Micro Systemation SIM Card Reader to acquire data from a populated SIM.

SIM 5343

The following scenarios were conducted using a T-Mobile SIM. Outgoing SMS messages were not populated, and Service Provider Name (SPN) was not allocated. No FPLMN entries were registered.

Basic Data: The following data was found and reported: IMSI, ICCID, Abbreviated Dialing Numbers (ADN), Last Numbers Dialed (LND) and active/deleted incoming SMS messages. Language Preference (LP) data was not found. (Below)

Location Data: The following LOCI data was found and reported: TMSI and LAI. However, the Location Update Status was incorrectly presented. GPRSLOCI data was not found. (Below)

EMS Data: Active/deleted incoming EMS Messages that exceed 160 characters or contain an embedded picture were found and reported. (Meet)

Foreign Language Data: ADN entries and SMS messages containing French language characters were found and reported. ADN entries and SMS messages containing Asian language characters were found, but the text characters were not presented correctly. (Below)

SIM 8778

The following scenarios were conducted using a Cingular SIM. Service Provider Name (SPN) was allocated but not activated. No FPLMN entries were registered.

Basic Data: The following data was found and reported: IMSI, ICCID, Abbreviated Dialing Numbers (ADN), Last Numbers Dialed (LND) and active/deleted incoming SMS messages. Language Preference (LP) data was not found. (Below)

Location Data: The following LOCI data was found and reported: TMSI and LAI. However, the Location Update Status was incorrectly presented. GPRSLOCI data was not found. (Below)

EMS Data: Active/deleted incoming EMS Messages that exceed 160 characters or contain an embedded picture were found and reported. (Meet)

Foreign Language Data: ADN entries and SMS messages containing French language characters were found and reported. ADN entries and SMS messages containing Asian language characters were found, but the text characters were not presented correctly. (Below)

SIM 1144

The following scenarios were conducted using an AT&T SIM. Service Provider Name (SPN) was not allocated. No FPLMN entries were registered.

168

Basic Data: The following data was found and reported: IMSI, ICCID, Abbreviated Dialing Numbers (ADN), Last Numbers Dialed (LND) and active/deleted SMS messages. However, the MNC portion of the IMSI, a three-digit value, was incorrectly decoded and reported. Language Preference (LP) data was not found. (Below)

Location Data: The following LOCI data was found and reported: TMSI and LAI. However, the Location Update Status was incorrectly presented. GPRSLOCI data was not found. (Below)

EMS Data: Active/deleted incoming EMS Messages that exceed 160 characters or contain an embedded picture were found and reported. (Meet)

Foreign Language Data: ADN entries and SMS messages containing French language characters were found and reported. ADN entries and SMS messages containing Asian language characters were found, but the text characters were not presented correctly . (Below)

Appendix P: Mobiledit! – External SIM Results

The scenarios were performed on a Forensic Recovery of Evidence Device (FRED) running Windows XP SP2. Mobiledit! version 1.95 was used with a PC/SC SIM Card Reader to acquire data from a populated SIM.

SIM 5343

The following scenarios were conducted using a T-Mobile SIM. Outgoing SMS messages were not populated, and Service Provider Name (SPN) was not allocated. No FPLMN entries were registered.

Basic Data: The following data was found and reported: IMSI, ICCID, Abbreviated Dialing Numbers (ADN), Last Numbers Dialed (LND) and active incoming SMS messages. The following data was not found: Language Preference (LP) and deleted SMS messages. (Below)

Location Data: No data was found. (Miss)

EMS Data: Active incoming EMS Messages that exceed 160 characters were found and reported. An EMS message containing an embedded picture was found. However, the image was not correctly decoded and presented, though the text was. Deleted EMS messages were not found. (Below)

Foreign Language Data: ADN entries and SMS messages containing French language characters were found and reported. However, ADN entries and SMS messages containing Asian language characters were not displayed correctly in the user interface, although the SMS messages appeared correctly in the generated .rtf report. (Below)

SIM 8778

The following scenarios were conducted using a Cingular SIM. Service Provider Name (SPN) was allocated but not activated. No FPLMN entries were registered.

Basic Data: The following data was found and reported: IMSI, ICCID, Abbreviated Dialing Numbers (ADN), Last Numbers Dialed (LND) and active incoming SMS messages. The following data was not found: Language Preference (LP), and deleted SMS messages. The initial portion of the IMSI was incorrectly appended to the reported ICCID. (Below)

Location Data: No data was found. (Miss)

EMS Data: Active incoming EMS Messages that exceed 160 characters were found and reported. An EMS message containing an embedded picture was found. However, the image was not correctly decoded and presented, though the text was. Deleted EMS messages were not found. (Below)

Foreign Language Data: ADN entries and SMS messages containing French language characters were found and reported. However, ADN entries and SMS messages containing

170

Asian language characters were not displayed correctly in the user interface, although the SMS messages appeared correctly in the generated .rtf report. (Below)

SIM 1144

The following scenarios were conducted using an AT&T SIM. Service Provider Name (SPN) was not allocated. No FPLMN entries were registered.

Basic Data: The following data was found and reported: IMSI, ICCID, Abbreviated Dialing Numbers (ADN), Last Numbers Dialed (LND) and active SMS messages. The following data was not found: Language Preference (LP) and deleted SMS messages. The initial portion of the IMSI was incorrectly appended to the reported ICCID. (Below)

Location Data: No data was found. (Miss)

EMS Data: Active incoming EMS Messages that exceed 160 characters were found and reported. An EMS message containing an embedded picture was found. However, the image was not correctly decoded and presented, though the text was. Deleted EMS messages were not found. (Below)

Foreign Language Data: ADN entries and SMS messages containing French language characters were found and reported. However, ADN entries and SMS messages containing Asian language characters were not displayed correctly in the user interface, although the SMS messages appeared correctly in the generated .rtf report. (Below)

Appendix Q: SIM Services

Table 25 provides a complete list of SIM Service Table (SST) entries for the SIMs used to conduct the SIM scenarios with the various tools. More information about the SST entries and SIM card details can be found in 3GPP TS 11.11, Specification of the Subscriber Identity Module - Mobile Equipment (SIM - ME) interface.[21]

Table 25: SIM Service Table Entries

SIM	Network	Services	Allocated	Activated
1144	ATT	CHV1 disable function	X	X
		Abbreviated Dialing Numbers (ADN)	X	X
		Fixed Dialing Numbers (FDN)	X	X
		Short Message Storage (SMS)	X	X
		Advice of Charge (AoC)	X	
		Capability Configuration Parameters (CCP)	X	X
		PLMN Selector	X	X
		Called Party Subaddress		
		MSISDN	X	X
		Extension1	X	X
		Extension2	X	X
		SMS Parameters	X	X
		Last Numbers Dialed (LND)	X	X
		Cell Broadcast Message Identifier	X	X
		Group Identifier Level 1		
		Group Identifier Level 2		
		Service Provider Name		
		Service Dialing Numbers	X	
		Extension 3	X	X
		RFU	X	
		VGCS Group Identifier List	X	X
		VBS Group Identifier List	X	X
		Enhanced Multi-Level Precedence and Preemption Service		
		Automatic Answer for eMLPP		
		Data download via SMS-CB		
		Data download via SMS-PP	X	X
		Menu selection	X	X
		Call control		
		Proactive SIM	X	X
		Cell Broadcast Message Identifier Ranges	X	X
		Barred Dialing Numbers	X	
		Extension 4		
		De-personalization Control Keys		
		Cooperative network list		
		Short Message Status Reports	X	X
		Network's indication of alerting in the MS		

---

[21] Current and previous versions of the specification can be found at http://www.3gpp.org/ftp/Specs/html-info/1111.htm

SIM	Network	Services	Allocated	Activated
		Mobile Originated Short Message Control by the SIM		
		GPRS	X	X
		Image (IMG)		
		SoLSA (Support of Local Service Area)		
		USSD string data object supported in Call Control		
		RUN AT COMMAND command	X	X
		User Controlled PLMN selector with Access Technology		
		Operator Controlled PLMN selector with Access Technology:		
		HPLMN selector with Access Technology		
		CPBCCH information		
		Investigation Scan		
		Extended Capability Configuration Parameters		
		MExE		
		RPLMN last used Access Technology		
		Service no. 51	X	X
		Service no. 52	X	X
		Service no. 53	X	
		Service no. 54	X	
		Service no. 55	X	
		Service no. 56		
		Service no. 57	X	X
		Service no. 58	X	X
		Service no. 59	X	X
		Service no. 60		
8778	Cingular	CHV1 disable function	X	X
		Abbreviated Dialing Numbers (ADN)	X	X
		Fixed Dialing Numbers (FDN)	X	X
		Short Message Storage (SMS)	X	X
		Advice of Charge (AoC)		
		Capability Configuration Parameters (CCP)		
		PLMN Selector	X	X
		Called Party Subaddress		
		MSISDN	X	X
		Extension1	X	X
		Extension2	X	X
		SMS Parameters	X	X
		Last Numbers Dialed (LND)	X	X
		Cell Broadcast Message Identifier	X	X
		Group Identifier Level 1	X	X
		Group Identifier Level 2	X	X
		Service Provider Name	X	
		Service Dialing Numbers	X	X
		Extension 3	X	X
		RFU		

SIM	Network	Services	Allocated	Activated
		VGCS Group Identifier List		
		VBS Group Identifier List		
		Enhanced Multi-Level Precedence and Preemption Service		
		Automatic Answer for eMLPP		
		Data download via SMS-CB		
		Data download via SMS-PP	X	X
		Menu selection	X	X
		Call control		
		Proactive SIM	X	X
		Cell Broadcast Message Identifier Ranges		
		Barred Dialing Numbers		
		Extension 4		
		De-personalization Control Keys		
		Cooperative network list		
		Short Message Status Reports		
		Network's indication of alerting in the MS		
		Mobile Originated Short Message Control by the SIM		
		GPRS	X	X
		Image (IMG)		
		SoLSA (Support of Local Service Area)		
		USSD string data object supported in Call Control		
		RUN AT COMMAND command		
		User Controlled PLMN selector with Access Technology		
		Operator Controlled PLMN selector with Access Technology:		
		HPLMN selector with Access Technology		
		CPBCCH information		
		Investigation Scan		
		Extended Capability Configuration Parameters		
		MExE		
		RPLMN last used Access Technology		
		Service no. 51	X	X
		Service no. 52	X	X
		Service no. 53		
		Service no. 54		
		Service no. 55		
		Service no. 56		
		Service no. 57		
		Service no. 58		
		Service no. 59		
		Service no. 60		
		CHV1 disable function	X	X
5343	T-Mobile	Abbreviated Dialing Numbers (ADN)	X	X
		Fixed Dialing Numbers (FDN)	X	X

174

SIM	Network	Services	Allocated	Activated
		Short Message Storage (SMS)	X	X
		Advice of Charge (AoC)		
		Capability Configuration Parameters (CCP)	X	X
		PLMN Selector	X	X
		Called Party Subaddress		
		MSISDN	X	X
		Extension1	X	X
		Extension2	X	X
		SMS Parameters	X	X
		Last Numbers Dialed (LND)	X	X
		Cell Broadcast Message Identifier	X	X
		Group Identifier Level 1		
		Group Identifier Level 2		
		Service Provider Name		
		Service Dialing Numbers		
		Extension 3		
		RFU		
		VGCS Group Identifier List		
		VBS Group Identifier List		
		Enhanced Multi-Level Precedence and Preemption Service		
		Automatic Answer for eMLPP		
		Data download via SMS-CB		
		Data download via SMS-PP	X	X
		Menu selection	X	X
		Call control		
		Proactive SIM	X	X
		Cell Broadcast Message Identifier Ranges		
		Barred Dialing Numbers		
		Extension 4		
		De-personalization Control Keys	X	X
		Cooperative network list		
		Short Message Status Reports	X	X
		Network's indication of alerting in the MS		
		Mobile Originated Short Message Control by the SIM		
		GPRS	X	X
		Image (IMG)		
		SoLSA (Support of Local Service Area)		
		USSD string data object supported in Call Control		
		RUN AT COMMAND command		
		User Controlled PLMN selector with Access Technology		
		Operator Controlled PLMN selector with Access Technology:		
		HPLMN selector with Access Technology		
		CPBCCH information		
		Investigation Scan		

SIM	Network	Services	Allocated	Activated
		Extended Capability Configuration Parameters		
		MExE		
		RPLMN last used Access Technology		
		Service no. 51	X	X
		Service no. 52	X	X
		Service no. 53	X	X
		Service no. 54		
		Service no. 55		
		Service no. 56		
		Service no. 57		
		Service no. 58		
		Service no. 59		
		Service no. 60		
		Service no. 61		
		Service no. 62		
		Service no. 63		
		Service no. 64		

Printed in Great Britain
by Amazon